Use of Discrete-Zone Monitoring Systems for Hydraulic Characterization of a Fractured-Rock Aquifer at the University of Connecticut Landfill, Storrs, Connecticut, 1999 to 2002

By Carole D. Johnson, Christopher S. Kochiss, and C.B. Dawson

In cooperation with the University of Connecticut

Water-Resources Investigations Report 03-4338

U.S. Department of the Interior
U.S. Geological Survey

U.S. Department of the Interior
Gale A. Norton, Secretary

U.S. Geological Survey
Charles G. Groat, Director

U.S. Geological Survey, Reston, Virginia: 2005
For sale by U.S. Geological Survey, Information Services
Box 25286, Denver Federal Center
Denver, CO 80225

For more information about the USGS and its products:
Telephone: 1-888-ASK-USGS
World Wide Web: http://www.usgs.gov/

Contents

Figures

Tables

Conversion Factors, Vertical Datum, and Abbreviations

Multiply	By	To obtain
inch (in.)	25.4	millimeter
foot (ft)	0.3048	meter
foot squared (ft^2)	0.09290	meter squared
acre	4,047	meter squared
gallon (gal)	0.003785	meter cubed
gallon per minute (gal/min)	3.7854	liters per minute
pounds force per square inch (psi)	6.89	kilopascal

Temperature in degrees Fahrenheit ($^\circ$F) can be converted to degrees Celsius ($^\circ$C) as follows:

$^\circ$C = 5/9 ($^\circ$F - 32)

Vertical datum: Vertical coordinate information is referenced to the North American Vertical Datum of 1988 (NAVD 88). For the UConn landfill study area, the change from National Geodetic Vertical Datum of 1929 (NGVD 29) to NAVD 88 is 0.830 ± 0.003 foot.

Transmissivity: Transmissivity is a measure of the ability of the aquifer to transmit water. The units of measurement are [(ft^3/d)/ft^2] ft, which represents a volumetric flow rate over a given area for a known thickness of aquifer. These terms simplify to units of ft^2/d, which will be used in this report.

Other abbreviations used in this report:

$^\circ$	degree
µg/L	microgram per liter
d	day
ft^2/d	foot (feet) squared per day
ft^2/s	foot (feet) squared per second
gal/d	gallon per day
gal/min/ft	gallon per minute per foot
(gal/min)^2	gallon per minute, squared
h	hour
min	minute
mS/m	millisiemen per meter
mV	millivolt
psi	pounds per square inch
s	second
yr	year

Use of Discrete-Zone Monitoring Systems for Hydraulic Characterization of a Fractured-Rock Aquifer at the University of Connecticut Landfill, Storrs, Connecticut, 1999 to 2002

By Carole D. Johnson, Christopher S. Kochiss, and C.B. Dawson

Abstract

The U.S. Geological Survey, in cooperation with the University of Connecticut, used a suite of hydraulic methods to characterize the hydrogeology of a fractured-rock aquifer near the former landfill and chemical-waste disposal pits at the University of Connecticut, Storrs, Connecticut. Multiple methods were used to determine head, driving potential, and transmissivity, including manual open-hole water-level and discrete-zone water-level measurements from 11 boreholes; continuous discrete-zone water-level measurements from 6 of the boreholes; estimated head and transmissivity for 11 boreholes using heat-pulse flowmeter profiles and pumping records; and differential head testing using a straddle-packer apparatus from 4 boreholes. These data were analyzed to identify and characterize relations between long-term water-level patterns and precipitation, topographic setting, contaminant distribution at the site, and a conceptual ground-water flow model.

Data collected using the heat-pulse flowmeter, the straddle-packer apparatus, and discrete-zone monitoring (DZM) systems helped to establish, refine, and verify a conceptual model of ground-water flow in the study area. Monitoring of DZM systems installed in 11 boreholes provided a method for long-term monitoring of hydraulic head and water quality of the aquifer at fracture zones of different depths. These data were used to help define the conceptual site model for ground-water flow and to determine and explain the distribution of contamination.

Hydrographs constructed for discretely isolated zones in the boreholes showed the magnitude of seasonal changes of water levels and driving potential in response to precipitation and drought. Heads in discrete zones and in different boreholes varied both in magnitude of response and in timing of response to precipitation. Water levels in open boreholes and in DZM systems showed a semi-diurnal pattern that coincides with gravimetric tidal plots generated for this area. No fluctuations that might indicate pumping were identified in the continuous water-level records. Lack of hydraulic response between boreholes during cross-hole testing in the area of the former chemical-waste disposal pits indicates poor hydraulic connection between the boreholes that were tested. In general, data indicated the presence of downward driving potentials in the recharge areas and in the area of the ground-water divide, and upward driving potentials in discharge areas north and south of the landfill.

The results of this study illustrate the importance of discrete-zone isolation and monitoring in fractured-rock aquifers to prevent cross contamination while permitting head measurements and water-quality sampling that can be used to identify and characterize contamination or pathways for contaminant migration in a fractured-rock aquifer. Without DZM systems installed in the boreholes, only open-hole heads can be measured. The open-hole heads may be misleading when determining potential flow directions at contamination sites, because they are a composite of the heads associated with each of the fractures intersecting the borehole. The flowmeter tool and straddle-packer apparatus are effective screening tools for generating a snapshot of the hydraulic conditions, including vertical flow, transmissivity, and heads; however, they cannot prevent flow and potential cross-contamination and cannot easily be used to monitor long-term conditions.

This work was conducted as part of a larger multidisciplinary investigation to characterize the nature and extent of contamination in the soil, surface water, and ground water in the overburden and fractured bedrock in the area of the landfill and former chemical-waste disposal pits near the University of Connecticut. The methods and hydraulic data presented in this report were used along with surface- and borehole-geophysical data and geochemical data to understand and characterize the ground-water flow in overburden and fractured bedrock; to assess possible chemical migration; to develop a site conceptual ground-water flow model; and to assess remediation alternatives.

Introduction

From 1999 through 2002, the U.S. Geological Survey (USGS), in cooperation with the University of Connecticut (UConn), used a suite of geophysical and hydraulic methods to characterize the hydrogeology of a fractured-rock aquifer near a former landfill and chemical-waste disposal pits at UConn in Storrs, Connecticut. The study demonstrated the collective use of surface- and borehole-geophysical methods to identify contamination or pathways for contaminant migration in a fractured-rock aquifer and highlighted the importance of isolating hydraulically active fractures to prevent cross-contamination while permitting discrete-interval head and water-quality sampling. To further characterize the hydrogeology of the site, the USGS installed discrete-zone monitoring (DZM) systems in 11 boreholes completed in the bedrock. The boreholes are about 125 ft deep and are completed with 2 to 5 discrete zones. Vertical flow under ambient conditions was detected in four wells using a heat-pulse flowmeter, which can measure flow rates as low as 0.01 ± 0.005 gal/min. The amount of measured vertical flow in these boreholes, which ranged from about 14 to 300 gal/d (Johnson and others, 2002), necessitated a DZM system to prevent possible cross contamination.

A DZM approach was used to characterize the ground-water flow and contaminant distribution in a fractured-rock aquifer. Because open-hole head data and water-quality samples from bedrock wells can be misleading, it was important to obtain discrete-interval data from the bedrock. Head data collected from discretely isolated zones can be used to help define the conceptual site model for ground-water flow, and water-quality data used in conjunction with the head data can be used to explain the distribution of contamination. The installation of DZM systems has the added benefit of minimizing the spread and dilution of contaminants, while enhancing the probability of collecting representative samples that can be used to help define the flow regime and contaminant distribution.

Purpose and Scope

This report describes the design and use of DZM systems installed in 11 boreholes completed in bedrock at the UConn landfill. All water-level data were collected from 2000 through 2002 and are summarized in this report. These data include: manual open-hole water-level and discrete-zone water-level measurements from 11 boreholes, continuous discrete-zone water-level measurements from 6 of the boreholes, estimated head and transmissivity for 11 boreholes using heat-pulse flowmeter profiles and pumping records, and the results of differential head testing using a straddle-packer apparatus from 4 boreholes. The report also provides a discussion of long-term water-level monitoring relative to precipitation, topographic setting, contaminant distribution at the site, and a conceptual ground-water flow model of the study site.

Hydrogeologic Setting

The UConn campus is in Storrs, Connecticut, in the northeastern part of the State (fig. 1). The UConn landfill is in the northwestern corner of the campus and covers about 15 acres. The study area occupies a north-south trending valley with highlands on the northeast and southwest. The study area is bounded on the east by a steep hill and on the west by locally minor topographic hills that are topographically higher than the land surface of the landfill. The landfill overlies a minor

ground-water divide that drains to the north and south along the axis of the valley. The surface runoff flows north through a wetland towards Cedar Swamp Brook and south towards Eagleville Brook through a seasonal drainage. Regional ground-water flow is inferred to follow the surface topography; however, the local flow and transport in bedrock follows fractures that may be oriented differently than the regional gradient. In this report, the term "UConn landfill study area" is used to describe the area shown in figure 1 and includes the landfill, the former chemical-waste disposal pits, and the area southwest of the landfill near the intersection of Hunting Lodge and North Eagleville Roads.

The bedrock that underlies the UConn landfill study area is folded, faulted, and fractured gneiss (Fahey and Pease, 1977). The bedrock aquifer is overlain by glacial till and other unconsolidated deposits, which range in thickness typically from 0 to about 20 ft and locally up to 50 ft.

A conceptual model of the ground-water flow system consists of hydraulically connected overburden and bedrock. Ground water originates as precipitation and infiltrates through the overburden to the water table and into the bedrock. Multiple sets of interconnected fractures provide pathways for water to move through the bedrock. Water moves from high head to low head through fractures. Ground-water flow generally follows the surface topography, and local flow systems discharge locally to streams and seeps. Boreholes outfitted with DZM systems were designed to "tap" into the aquifer and measure the hydraulic head from discrete zones in the bedrock aquifer to determine the difference in head between zones and the direction of driving potential between the zones. These data were used to identify flowpaths in the study area and were compared to water-quality data collected from discrete intervals in the bedrock and overburden.

Discrete-Zone Head Monitoring

Hydraulic head, which is an indicator of the total energy available to move ground water through an aquifer, is controlled by properties of the aquifer, including porosity, fracture connectivity, transmissivity, and thickness of the aquifer. Because ground water flows from higher head to lower head, the hydraulic head can be used to assess the general directions of flow in the aquifer. The actual flowpaths in fractured-rock aquifers are along a network of connected fractures, which may be far more complex than the generalized flow directions indicated by the head measurements.

Monitoring hydraulic head is an important part of many hydrogeologic investigations, because head values are used to determine the rates and directions of ground-water flow. In unconfined aquifers, open-hole water levels commonly are used to establish meaningful conceptual models of ground-water flow. In heterogeneous fractured-rock aquifers, the hydraulic head can be different for each fracture zone. This results in an open-hole water level that is actually an integrated head derived from the transmissivity-weighted heads associated with all of

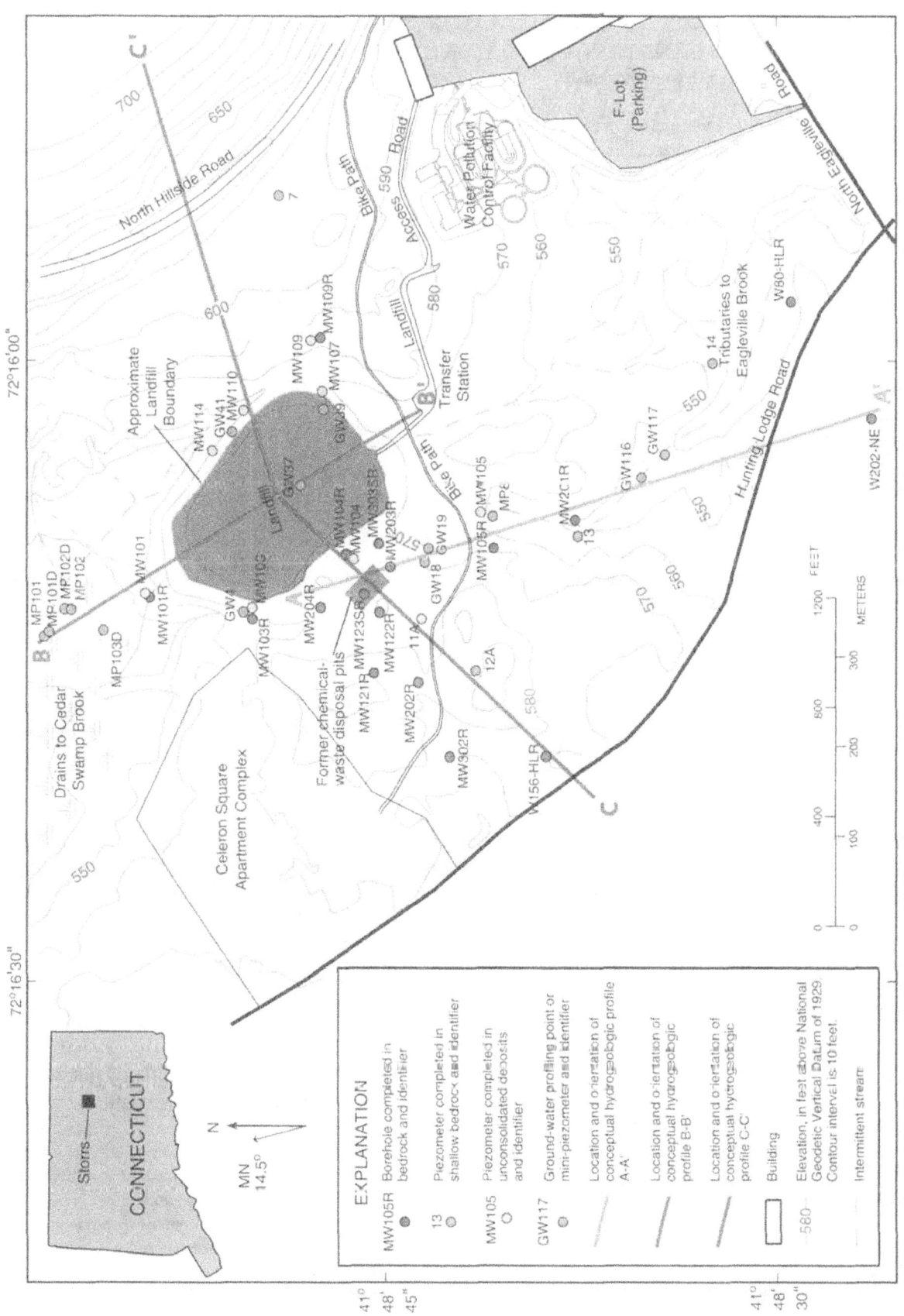

Figure 1. Location of boreholes, ground-water profiling points, piezometers, and conceptual hydrogeologic profiles in the UConn landfill study area, Storrs, Connecticut.

Base from Haley and Aldrich, Inc. and others. 2000. 1:2,400 scale

the heads of the individual fractures intersecting the borehole. As a result, open-hole water levels can be misleading and may not be reliable for determining the rates and directions of ground-water flow and for defining a conceptual model of ground-water flow.

In heterogeneous fractured-rock aquifers, a DZM system is necessary to determine the spatial distribution and temporal variation of hydraulic head. The information provided by DZM systems can be used to determine vertical and horizontal hydraulic gradients, to identify connections between transmissive zones within the aquifer, and to establish the extent of connection between the fractured-rock and unconsolidated surficial aquifers.

Use of a DZM system can avoid some of the problems associated with water-quality sampling in bedrock wells, because DZM systems allow for sampling from isolated sections of the boreholes. Water samples from open boreholes can be difficult to interpret in the presence of vertical flow or multiple transmissive zones. Water-quality samples obtained from open holes where vertical flow has occurred can be ambiguous, misleading, or completely erroneous (Shapiro, 2002). In the presence of vertical flow, sampling at a low flow rate adjacent to a fracture does not insure that the water sample will come only from that fracture. That is, much or all of the sample water would be derived from any inflowing fractures in the borehole. In addition, if a fracture that had been receiving water from one or more inflowing fractures was isolated with straddle packers just prior to sampling, then the water sample would likely be representative of the inflowing fracture (Shapiro, 2002). Installation of a DZM system soon after drilling minimizes vertical flow and ensures the collection of a water-quality sample from a discretely isolated zone. Installation of a long-term DZM system may allow the aquifer to approach or return to water-quality conditions present prior to drilling the new borehole; however a long time and active purging may be required to remove the effects of cross contamination (Sterling, 1999; Elci and others, 2001). Water-quality samples from DZM systems were used to characterize the spatial distribution of ground-water quality and contamination in the study area.

In addition to providing a means to monitor head and water quality for individual fractures or fracture zones, DZM systems perform the equally important role of preventing cross contamination between fractures. Hydraulic heads associated with individual fracture zones vary within a single well, creating the potential for vertical flow within the well. At contaminated sites, vertical flow can spread contamination through the open hole from zones of higher head to zones of lower head, which may previously have been uncontaminated (Williams and Conger, 1990; Parker and Cherry, 1999). Thus, open-hole monitoring may result in further contamination of the aquifer.

The open-hole hydraulic head collected in a borehole is an integrated head that is a transmissivity-weighted average of the heads of all fractures that intersect the borehole. Thus the head of a highly transmissive fracture will dominate the open-hole hydraulic head. Open-borehole hydraulic heads cannot be used to accurately assess the hydraulic head in individual fractures

that intersect the borehole, unless the borehole intersects only one fracture or all of the fractures have the same head. In this investigation, three methods were used to determine the hydraulic head of individual fractures, vertical head differences, and driving potential between fractures.

Acknowledgments

The authors gratefully acknowledge the technical guidance and cooperation of Susan Soloyanis of Mitretek Systems and James Pietrzak of UConn. The technical insight and assistance of F.P. Haeni, Remo Mondazzi, Allen Shapiro and Frederick Day-Lewis of the USGS were essential to completion of this work. The cooperation and assistance of John Kastrinos, Elida Danaher, Steve Brousseau, and Mike Alfieri of Haley and Aldrich, Inc. were greatly appreciated. Also the authors acknowledge the many USGS personnel who provided assistance in the installation of the DZM systems and with the collection and analysis of data: Marcel Belaval, Ellen Douglas, Peter Joesten, Grady O'Brien, and Eric White. The authors would like to thank Jim Pianosi of Solinst Canada, Ltd., and Marc Zabowa and Jan Mathews of Druck, Inc. for their assistance in maximizing the functionality of the DZM systems and water-level monitoring equipment.

Methods of Investigation

For this investigation, hydraulic heads were determined using three different methods. One method compared the results of observed heat-pulse flowmeter measurements with numerically simulated heat-pulse flowmeter profiles using estimated fracture transmissivity and head for individual fractures (Paillet, 1998, 2000). The other two methods involved isolating individual fractures or fracture zones to measure the hydraulic head. Differential head tests were conducted using a straddle-packer apparatus over a period of minutes to hours. In addition, DZM systems were installed for long-term water-level monitoring. Water levels were monitored in boreholes MW101R, MW103R, MW104R, MW105R, MW109R, MW121R and MW122R for 2 years and in boreholes MW201R, MW202R, MW203R, and MW204R for about 1.5 years (fig. 1).

Estimation of Transmissivity and Hydraulic Head of Fractures with Heat-Pulse Flowmeter Data

Using the computer program of Paillet (2000), the hydraulic head and transmissivity were determined for each hydraulically active zone. The method requires two non-zero flow regimes, including either flow data for ambient and one pumping condition or flow data for two different pumping conditions, and the difference in water level between the two regimes. The program simulates two flow regimes as a function of transmissivity and hydraulic head from all productive zones intersecting the borehole. Model input requires information on the location of the transmissive fractures, ambient open-hole water-level data, quasi steady-state water-level data under pumping conditions, heat-pulse flowmeter data under two flow regimes, and initial estimates of transmissivity and head for each zone. The model produces ambient and pumped profiles that are qualitatively matched to the observed flow profiles. The simulated flow profiles were compared to measured flow profiles, and the head and transmissivity parameters were iteratively varied until a unique solution was obtained for the transmissivity and head of each fracture. The model also provides a sum-of-squares (SS) error in gallons per minute squared to assess the fit of the predicted flow profile to the measured profile, and hence the validity of the solution.

Differential-Head Tests with the Bedrock-Aquifer Transportable Testing Tool (BAT³)

Discrete fractured intervals of selected boreholes temporarily were isolated with an inflatable straddle-packer apparatus to conduct fluid-withdrawal tests and differential head tests. A straddle-packer apparatus, called the BAT³ (Bedrock-Aquifer Transportable Testing Tool), was used to obtain the samples and monitor the heads during pumping (Shapiro, 2001). A detailed description of the methods and the results of these tests are provided by Johnson and others (2005). Results of water-quality sampling were summarized by Haley and Aldrich, Inc. and others (2002). Two inflatable packers separated by 6.5 ft were used to isolate the test zone from the zones above and below the test interval. Pressure-sensitive transducers measured the heads in, above, and below the test zone. A submersible, variable-rate pump located between the two packers was used to directly sample the fractures that intersect the test zone.

The packers were lowered into position, inflated, and left for the heads to equilibrate. Zones of lower transmissivity require longer periods of time to equilibrate. For zones that contained fractures that were not identified as hydraulically active with the heat-pulse flowmeter, the packers were left overnight to equilibrate. Once the heads reached a quasi steady state, the head data were collected and the packers were moved to another test interval. Thus, the BAT³ testing provided a snapshot of the head and water-quality conditions at the time of collection.

The BAT³ system suspended in the air prior to lowering it into borehole MW203R is shown in figure 2. The photograph shows the locations of the packers (A and B on fig. 2), the housings that contain pressure transducers for monitoring the hydraulic head in the test zone (C on fig. 2) and in the lower zone (D on fig. 2), and the pump for obtaining samples from the test zone (E on fig. 2). After the apparatus was in place, a third transducer was lowered into the top of the borehole to monitor the water level in the upper zone (not shown in fig. 2).

Relative to the transducers used in the long-term DZM program, the BAT³ equipment used higher-pressure transducers (50, 100, and 200 psi), which were designed to accommodate submergence at greater depths. In general, there is a trade-

off between transducer resolution and maximum submergence; thus, these higher-pressure transducers have lower resolution than the transducers used in the long-term monitoring program. The upper zone of the BAT³ apparatus was monitored with a 10-psi transducer that had a resolution of ± 0.06 percent full scale of the transducer, which is equivalent to 0.014 ft. The middle zone was monitored with a 100-psi transducer, which measured at a resolution of 0.139 ft, and the lowermost zone used a 200-psi transducer, which has a resolution of 0.277 ft. The transducers used to monitor the middle and lower zones were designed to accommodate and measure deep water. The upper transducer was lowered into the water to a depth approximately 10 ft below the water level at the time of submergence. Small magnitude differential heads less than the resolution of the transducers used in the BAT³ cannot be adequately measured with this equipment.

Long-Term Water-Level Monitoring with a Discrete-Zone Monitoring System

The conceptual model for ground-water flow at a site needs to reflect the change of flow with time. Heat-pulse flowmeter and BAT³ data can only be used to determine relative hydraulic heads of fracture zones within a borehole at specific times. In order to develop a conceptual model that more accurately represents the flow regime of a site over time, continuous measurements of hydraulic head for individual fracture zones in boreholes across the site must be made. Installation of a DZM system can address this need for long-term measurements while simultaneously preventing cross contamination that can occur between fractures in contaminated zones in an open hole. A DZM system was designed and implemented at the UConn landfill for these purposes. The DZM system allowed for collection of manual and automated water-level measurements in discrete zones within boreholes across the study area, including testing to assess cross connections between boreholes. The design, installation, and implementation of this DZM system are described in this section.

Design and Installation of the Discrete-Zone Monitoring System for the UConn Landfill

The results of a geophysical investigation and hydraulic testing were used to design a DZM network in the fractured rock aquifer (Johnson and others, 2005). Data from geophysical logs were used to determine (1) the locations of fractures; (2) the presence of vertical flow in the borehole; (3) the location of hydraulically active fractures sufficiently transmissive for pumping; (4) the location of smooth, unfractured portions of the borehole suitable for packer placement; and (5) locations for monitoring head and sampling water quality. In some cases the packers were placed close together to limit the amount of water that would have to be purged from the monitoring zone prior to sampling.

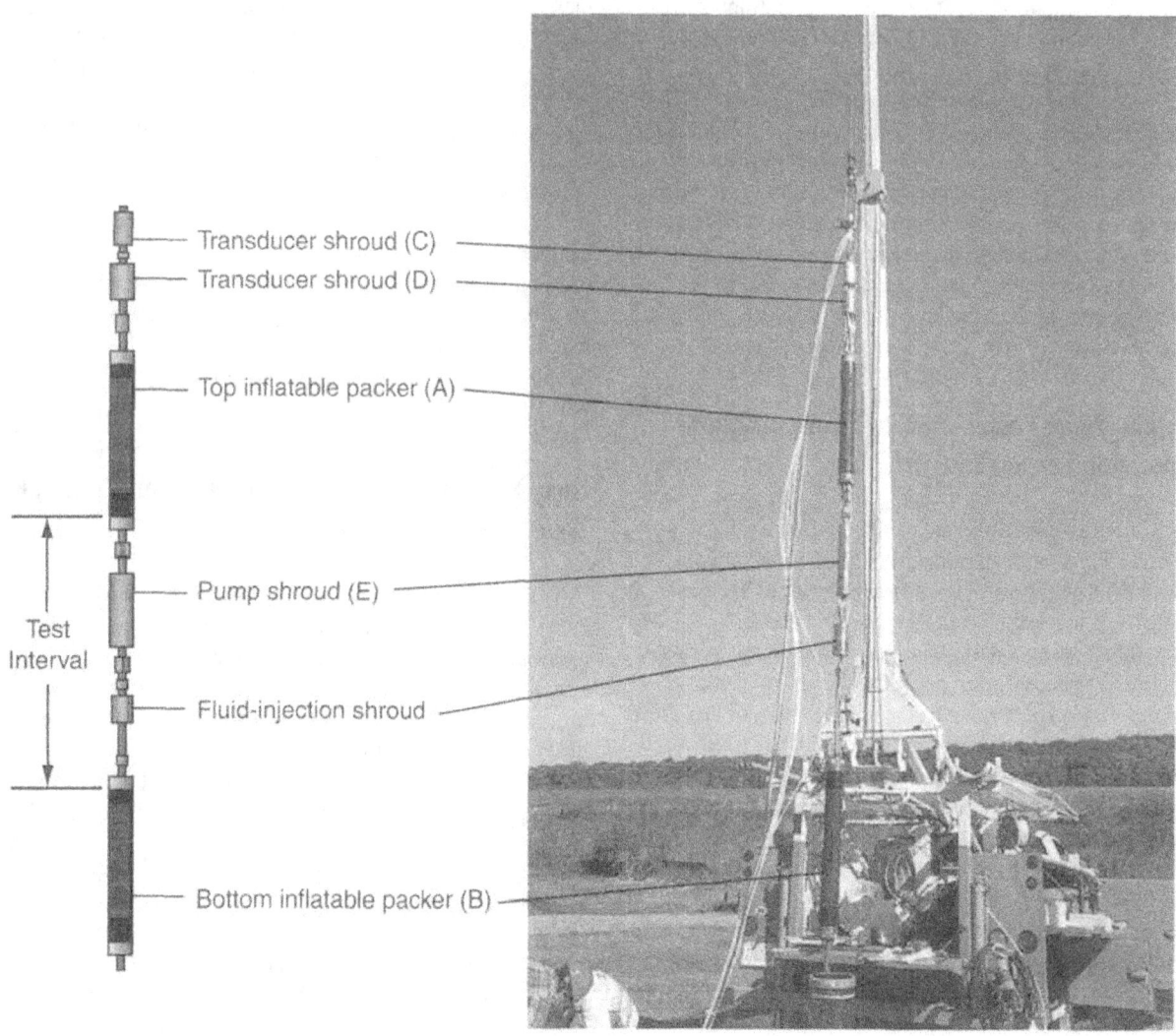

Figure 2. Schematic diagram and photograph of the multifunction bedrock-aquifer transportable testing tool (BAT^3).

Semi-permanent DZM systems were constructed using continuous multi-channel tubing (CMT) and an inflatable packer system commercially available from Solinst Canada. The DZM system, shown in figure 3, consists of a 1.7-in-diameter high-density polyethylene tube, divided into seven channels, and packers for isolating zones in the borehole. The CMT extends from the land surface to the lowermost zone isolated by packers. Packers were placed at selected locations along the tube, and individual channels were punctured at a specific depth with four 0.44-in-diameter holes to provide a sample port (fig. 3B). The ports were covered with stainless steel mesh to prevent particulate inflow into the channel. Below each port, the chan-

nel was sealed with polyethylene to hydraulically isolate the interval. This design allowed water to enter a channel in the tubing and rise to the head of the isolated zone. Another hole was drilled below the seal to allow air to escape from the channel and water to flow through the channel facilitating the emplacement of the packer system. Without this "flow-through" hole, the trapped air makes the entire CMT system too buoyant to install. Figure 3B shows preparation of the CMT system prior to installation.

The lowermost zone, which is bounded on the lower end by the bottom of the borehole, is connected to the center channel of the CMT. Packers (fig. 3C) located above and below each

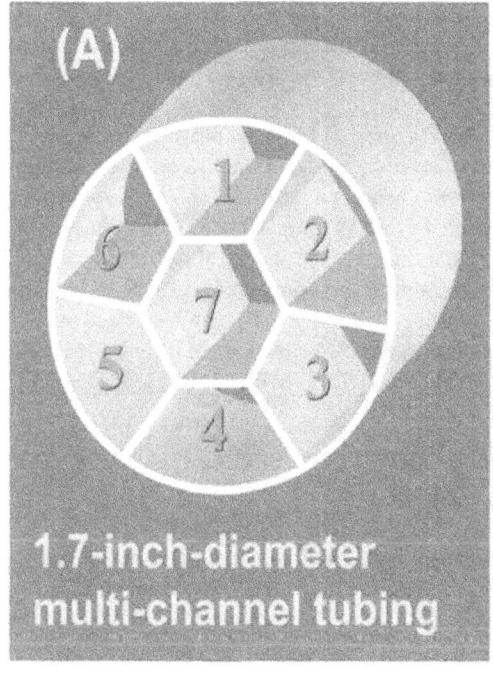

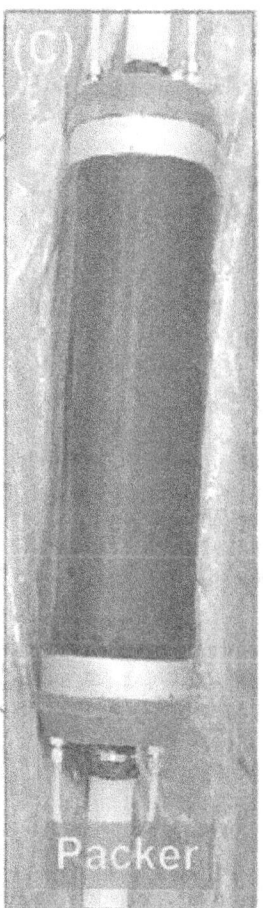

Figure 3. Discrete-zone monitoring system.
  (A) Center tubing showing multiple channels.
  (B) Packer array showing packer and port locations.
  (C) Inflatable 6-inch-diameter packer.

port were inflated with water to isolate each monitoring zone. With this design, the water level in each channel equilibrates to the hydraulic head in the isolated zone. An example of the DZM completion and the resultant hydrograph is shown in figure 4.

A comparison of the heads in the different zones indicates a driving potential between zones. In this example, the lowermost zone has the highest head and the uppermost zone has the lowest head, so the driving potential is upward. Water levels and water-quality samples can be obtained from the top of the well. The uppermost zone can be sampled outside the CMT tube, above the top packer.

The packers and tubing were assembled at land surface and then manually lowered into the borehole. The packers were then inflated with deionized water injected through 0.25-in-diameter tubing, and the air in the packers was displaced through another 0.25-in-diameter line vented above the land surface. The air-discharge line was then closed, and the packers were pressurized to about 20 psi above ambient pressure. A gage at the top of the well was used to monitor packer inflation over time. The packers typically lost some pressure over time and had to be re-inflated to 20 psi above ambient pressure about once every other month.

Unlike the DZM systems installed in the other boreholes, the 4.5-in-diameter packers that were used in MW121R and MW109R have only one inflation line. The injected water fills the packers through this line and the air has to be displaced up the same line. The DZM systems constructed with the smaller diameter packers were otherwise similar and were installed using the same process.

Although CMT systems could be used to monitor as many as eight individual zones (seven through the channels plus the uppermost zone outside the CMT), the DZMs for this investigation were designed to have at least two channels in each transmissive zone. This design allowed one channel to be used for continuous water-level monitoring while the other channel for the same zone could be used to check measurements and for water-quality sampling. While in most cases two channels were dedicated to each zone, in a few cases only one channel was used to monitor the zone.

The CMT design used in this DZM system limits sampling to the top of the well with a narrow-diameter pump that can access the isolated zones from the top of the well. A peristaltic pump was used to sample most zones. In cases where the water level was not within the depth-of-suction lift of a peristaltic pump, then a foot-valve pump was used to sample the isolated zones.

The boreholes in the UConn landfill study area are described in table 1, and the details of the DZM system completions are given in table 2.

Open-Hole Head Measurements

Open-hole head measurements were collected in the bedrock wells prior to the installation of DZM systems. In some cases, a comparison of the open-hole head with the DZM sys-

tem heads provided qualitative information about the relative transmissivity of the zones.

In addition, open-hole head measurements were collected in monitoring well MW302R, which did not have a DZM system installed. Well MW302R was monitored using a downhole transducer and data logger (Level Logger) manufactured by Solinst Canada. The transducer was not vented to atmospheric conditions, but was corrected for atmospheric pressure using pressure data collected by a second transducer. The resolution of the head measurements is ± 0.01 ft. Water levels were measured and recorded every 15 min, and over 1-min intervals for selected periods of time. The resolution of the transducer used for measuring the atmospheric pressure is ± 0.007 ft. In addition, manual measurements were collected in selected open-hole piezometers and shallow bedrock wells including MW101, MW103, MW104, MW109, 13, MW123SR, and MW303SR.

Manual Water-Level Measurements in Discrete-Zone Monitoring Systems

For this study, water levels were measured periodically, typically every 2 weeks to 1 month. These manual measurements were collected with an electric water-level meter in feet below the top of casing. The water levels were converted to elevation in feet above NAVD 88. The resolution of the water-level meter used in this investigation is ± 0.01 ft. The packer inflation pressure was monitored concurrently with water-level measurements.

Continuous Water-Level Monitoring Systems

Continuous water-level monitoring systems were installed to identify natural fluctuations and to identify possible human-induced stresses that could not be seen in bi-weekly measurements. Pressure transducers, which measure the pressure above a sensor and convert the pressure to voltage, were placed approximately 10-15 ft below the water surface at the time of installation, and measurements were recorded on a data logger at the well head. The data loggers were programmed to measure and record water levels every 15 min. Six boreholes were instrumented to obtain continuous head data in all the monitoring zones where check measurements could be maintained. All transducers were calibrated individually using periodic manual measurements. Transducers were calibrated over a range of submergences, and manual measurements were made to check the transducer data.

Transducers small enough to fit into the CMT channels were used to measure the heads in isolated zones. Druck PDCR35, 10-psi, strain-gage transducers were placed in individual channels in the DZM system. These 0.39-in-diameter transducers were attached to vented cables, which when kept dry, allowed the transducer to "breathe." Thus, the pressure readings did not need to be corrected for the ambient atmospheric pressure. The transducers were connected to and controlled by either a Campbell CR10X or CR510 data logger.

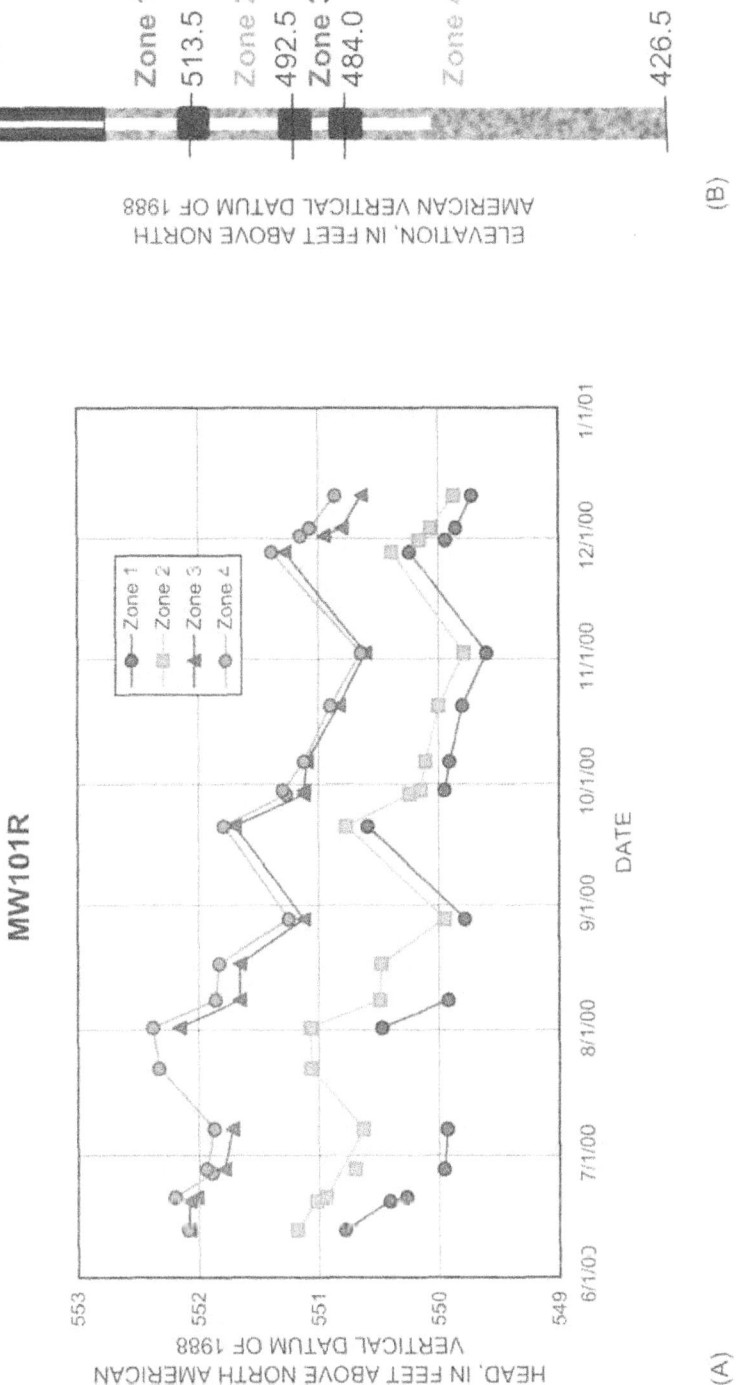

Figure 4. (A) Hydraulic heads in four isolated zones in bedrock borehole MW101R, UConn landfill, Storrs, Connecticut. Gaps in the record indicate missing or erroneous data that were omitted (B) Discrete-zone monitoring system completion in MW101R.

Table 1. Description of boreholes in the UConn landfill study area, Storrs, Connecticut.

[Height of measuring point is relative to land surface. Elevation is based on NAVD 88. All length, depth, and elevation measurements are in feet. BOC, bottom of casing; TOC, top of casing (steel); CMT, continuous multi-channel tubing; ~, indicates approximation; --, no data or not applicable]

Borehole	Date drilled	Date DZM installed	Diameter	Elevation of land surface	[1]Measuring point elevation	[1]Height of measuring point	Total depth from TOC	Depth to BOC from TOC	Location	Height of CMT above TOC	[2]Elevation of top of central tubing	[3]Height of 0.5-in pipe above TOC	Elevation of 0.5-in pipe
MW101R	8/24/99	6/14/00	0.5	552.8	553.53	0.7	127	23.0	N of landfill	0.39	553.92	--	--
MW103R	8/20/99	9/13/00	0.5	570.0	571.00	1.0	130	26.5	NW of landfill	0.40	571.40	0.37	571.37
MW104R	8/17/99	9/18/00	0.5	572.8	575.46	2.7	128	18.0	W of landfill	0.25	575.71	--	--
MW105R	7/23/99	8/09/00	0.5	563.5	565.79	2.4	125	12.0	SW of landfill	0.12	565.91	0.03	565.82
MW109R	7/02/99	5/04/00	0.4	580.1	582.28	2.2	126	20.0	E of landfill	0.36	582.64	0.28	582.56
MW121R	7/15/99	4/18/00	0.4	587.0	588.82	1.8	130	11.0	W of landfill	0.38	589.20	--	--
MW122R	7/22/99	7/27/00	0.5	578.8	581.40	2.4	127	11.5	W of landfill	0.36	581.76	--	--
MW201R	6/22/00	2/23/01	0.5	552.5	554.22	1.7	126	15.2	SW of landfill	0.46	554.68	0.32	554.54
MW202R	6/20/00	3/20/01	0.5	580.3	582.33	2.0	127	13.5	W of landfill	0.51	582.84	--	--
MW203R	6/21/00	3/29/01	0.5	575.5	576.90	1.4	125	14.4	W of landfill	0.48	577.38	--	--
MW204R	6/21/00	3/29/01	0.5	573.3	575.34	2.0	128	13.1	W of landfill	0.48	575.82	--	--
MW302R	12/20/00	N/A	[4]0.5	575.7	577.20	1.5	[5]303	127	W of landfill	--	--	--	--

[1] All measurements in this report are referenced to the measuring point, which is the top of steel casing. These measuring points were surveyed to within 0.01 ft of NAVD 88.

[2] Top of multiple-channel tubing.

[3] In selected boreholes, a 0.5-in-diameter casing was added to create another measurement channel used for head measurements in the uppermost zone.

[4] MW302R has 0.66-ft-diameter polyvinyl chloride casing and a 0.5-ft-diameter open hole.

[5] In April 2002, the total depth of MW302R was shortened to ~275 ft below top of casing.

Table 2. Discrete-zone monitoring systems installed in boreholes at the UConn landfill study area, Storrs, Connecticut.

[Height of measuring point is relative to land surface. Elevation is based on NAVD 88. All length, depth, and elevation measurements are in feet. MP, measuring point on steel casing; BOC, bottom of casing; BOH, bottom of borehole from top of casing; --, no data or not applicable; Z, zone; P, packer; DZM, discrete-zone monitoring; ~, indicates approximation]

Borehole	Elevation of MP on steel casing	Depth to BOC and BOH	Zone number	Depth to top of packer	Depth to bottom of packer	Elevation of top of zone and packer	Elevation of bottom of zone and packer	[1]DZM channels	[2]Water-quality sampling channels	[3]Transducer channel	Manual water-level measurement channel
MW101R	553.53	23	Z1			530.5	514.5	0, 1	--	1	0
			P	39.0	41.0	514.5	512.5				
			Z2			512.5	493.5	2, 3	2	3	2
			P	60.0	62.0	493.5	491.5				
			Z3			491.5	485.0	4, 5	4	5	4
			P	68.5	70.5	485.0	483.0				
		127	Z4			483.0	426.5	6, 7	--	--	6
MW103R	571.00	26.5	Z1			544.5	524.0	0, ½	0	½	0
			P	47.0	49.0	524.0	522.0				
			Z2			522.0	495.0	1, 2	--	2	1
			P	76.0	78.0	495.0	493.0				
			Z3			493.0	483.0	3, 4	3	4	3
			P	88.0	90.0	483.0	481.0				
			Z4			481.0	460.0	5	--	none	5
			P	111.0	113.0	460.0	458.0				
		130	Z5			458.0	440.0	6, 7	--	6	7
MW104R	575.46	8	Z1			557.5	539.5	0, 1	0	1	0
			P	36.0	38.0	539.5	537.5				
			Z2			537.5	503.5	2	--	none	2
			P	72.0	74.0	503.5	501.5				
			Z3			501.5	491.5	3, 4	--	4	3
			P	84.0	86.0	491.5	489.5				
			Z4			489.5	472.5	5	--	none	5
			P	103.0	105.0	472.5	470.5				
		128	Z5			470.5	448.5	6, 7	6	7	6

Table 2. Discrete-zone monitoring systems installed in boreholes at the UConn landfill study area, Storrs, Connecticut.—Continued

[Height of measuring point is relative to land surface. Elevation is based on NAVD 88. All length, depth, and elevation measurements are in feet. MP, measuring point on steel casing; BOC, bottom of casing; BOH, bottom of borehole from top of casing; --, no data or not applicable; Z, zone; P, packer; DZM, discrete-zone monitoring; ~, indicates approximation]

Borehole	Elevation of MP on steel casing	Depth to BOC and BOH	Zone number	Depth to top of packer	Depth to bottom of packer	Elevation of top of zone and packer	Elevation of bottom of zone and packer	[1]DZM channels	[2]Water-quality sampling channels	[3]Transducer channel	Manual water-level measurement channel
MW105R	565.79	12	Z1			553.8	511.3	0,½	--	½	0
			P	54.5	56.5	511.3	509.3				
			Z2			509.3	495.8	1,2	--	1	2
			P	70.0	72.0	495.8	493.8				
			Z3			493.8	485.8	3,4	3	4	3
			P	80.0	82.0	485.8	483.8				
			Z4			483.8	459.8	5	--	none	5
			P	106.0	108.0	459.8	457.8				
		125	Z5			457.8	438.8	6,7	6	7	6
MW109R	582.28	20	Z1			562.3	547.3	0,½	--	0	½
			P	35.0	37.0	547.3	545.3				
			Z2			545.3	533.3	2,3	--	2	3
			P	49.0	51.0	533.3	531.3				
			Z3			531.3	514.3	4,5	5	4	5
			P	68.0	70.0	514.3	512.3				
			Z4			512.3	489.3	1,6	1	6	1
			P	93.0	95.0	489.3	487.3				
		126	Z5			487.3	455.3	7	--	none	7
MW121R	588.82	11	Z1			577.8	533.3	0,1,2,3	0 or 1	none	2
			P	55.5	57.5	533.3	531.3				
		130	Z2			531.3	461.8	4,5,6	5	none	4
MW122R	581.40	11.5	Z1			569.9	522.9	0,1,2	0 or 1	2	0,1
			P	58.5	60.5	522.9	520.9				
			Z2			520.9	511.4	3,4,5	4	3	4,5
			P	70.0	72.0	511.4	509.4				
		127	Z3			509.4	454.4	6,7	--	none	6

Table 2. Discrete-zone monitoring systems installed in boreholes at the UConn landfill study area, Storrs, Connecticut.—Continued

[Height of measuring point is relative to land surface. Elevation is based on NAVD 88. All length, depth, and elevation measurements are in feet. MP, measuring point on steel casing; BOC, bottom of casing; BOH, bottom of borehole from top of casing; --, no data or not applicable; Z, zone; P, packer; DZM, discrete-zone monitoring; ~, indicates approximation]

Borehole	Elevation of MP on steel casing	Depth to BOC and BOH	Zone number	Depth to top of packer	Depth to bottom of packer	Elevation of top of zone and packer	Elevation of bottom of zone and packer	DZM channels [1]	Water-quality sampling channels [2]	Transducer channel [3]	Manual water-level measurement channel
MW201R	554.22	15.2	Z1			539.0	505.2	0, 1, ½	0 or 1	none	½
			P	49.0	51.0	505.2	503.2				
			Z2			503.2	485.2	2, 3	2	none	3
			P	69.0	71.0	485.2	483.2				
			Z3			483.2	465.2	4, 5	--	none	5
			P	89.0	91.0	465.2	463.2				
		125	Z4			463.2	428.6	6, 7	--	none	7
MW202R	582.33	13.5	Z1			568.8	545.3	0, 1, 2	0 or 1	none	2
			P	37.0	39.0	545.3	543.3				
			Z2			543.3	483.3	3, 4	--	none	4
			P	99.0	101.0	483.3	481.3				
		127	Z3			481.3	455.4	5, 6, 7	5 or 6	none	7
MW203R	576.90	14.4	Z1			562.5	533.9	0, 1, 2	0 or 1	none	2
			P	43.0	45.0	533.9	531.9				
			Z2			531.9	488.9	3, 4, 5	--	none	4
			P	88.0	90.0	488.9	486.9				
		125	Z3			486.9	451.9	6, 7	--	none	6
MW204R	575.34	13.1	Z1			562.2	545.3	0, 1, 2	0 or 1	none	2
			P	30.0	32.0	545.3	543.3				
			Z2			543.3	507.3	3, 4, 5	--	none	4
			P	68.0	70.0	507.3	505.3				
		128	Z3			505.3	447.6	6, 7	--	none	6

[1] DZM channel 0 indicates a measurement taken within the 6-in-diameter casing.

[2] Water-quality sampling channel 0 indicates sample collected outside multi-channel tubing.

[3] Transducer channel ½ indicates measurement was taken from within a ½-in-diameter tube that was suspended in the upper zone.

The pressure readings were recorded in millivolts. The manufacturer's suggested operating range is 10 psi for these transducers, which is equivalent to a maximum pressure head of 23 ft of water. Transducers were set approximately 10 ft below the estimated average water level. In some cases the transducers were not able to reach this target depth, because the transducers did not slide down the channels. This problem was attributed to small constrictions caused by centralizers placed outside the CMT at the time of installation or because of small irregularities in the channel sizes. Table 3 describes the transducer equipment used in each borehole, the set depth of the transducer, and the date of installation.

Table 3. Description of transducer equipment used in discrete-zone monitoring systems at the UConn landfill study area, Storrs, Connecticut, 2000 to 2002.

[DZM, discrete-zone monitoring]

Borehole	DZM zone	[1]DZM channel with transducer	Set depth, in feet	Transducer serial number	Installation date
MW101R	1	1	14.7	1357075	9/5/01
MW101R	2	3	15.0	1357079	9/5/01
MW101R	3	5	15.0	1353519	9/5/01
MW103R	1	0	20.0	1353525	8/29/01
MW103R	2	2	20.0	1353521	8/29/01
MW103R	3	4	20.0	1353526	8/29/01
MW103R	5	6	25.0	1357080	8/29/01
MW104R	1	1	30.0	1354710	8/29/01
MW104R	3	4	35.0	1354714	8/29/01
MW104R	5	7	38.0	1354721	8/29/01
MW105R	1	½	20.0	1354711	8/30/01
MW105R	2	4	25.0	1354703	8/30/01
MW105R	3	6	25.0	1354702	8/30/01
MW105R	5	7	19.0	1354706	9/7/01
MW122R	1	2	30.0	1354704	8/30/01
MW122R	2	3	32.0	1354700	8/30/01
MW109R	1	0	30.2	1357074	9/7/01
MW109R	2	2	30.2	1357077	9/7/01
MW109R	3	4	30.2	1357076	9/7/01
MW109R	4	6	30.2	1353522	9/7/01

[1]DZM channel ½ indicates measurement was taken from within a ½-in-diameter tube that was suspended in the upper zone.

Although the transducers are designed for temperature stability, measurement repeatability, and linearity of the output, check measurements were made to adjust for any electronic drift or change in recorded signal that was not associated with a real change in the water level. Transducers were calibrated at the time of installation and again during the period of record. The measured voltage output was correlated to submergence over a range of submergences (fig. 5). The water levels in each channel were measured with an electric water-level meter. Typically, the transducers were set at their target depths in channels in the DZM systems, and the water levels were allowed to equilibrate. During the calibration process, the data loggers recorded pressure readings, in millivolts, every minute. After the water levels equilibrated, the transducers were raised to 1 ft above the starting point. The process was repeated so that several depths of submergences were recorded. A regression analysis was performed on the pressure data and measured water levels. The slope and offset of the regression line were used to convert millivolt readings to water levels. Results of the regression analyses, including the correlation coefficient, are included in appendix 1.

Each data logger was set to collect data at 15-min intervals. Continuous digital data were downloaded directly to a computer and subsequently processed, reviewed, and plotted for analysis. Data conversions are shown in figure 5. Processing included the following seven steps:

1. Using calibrations established for each transducer, the pressure measurements in millivolts were converted to submergence or height of water above the transducers in feet (A in fig. 5; A is depth of transducer, B is depth of submergence).
2. The submergence was converted to depth to water in feet below the measuring point (C in fig. 5).
3. The depth to water was converted to the elevation of head in feet above NAVD 88 (E in fig. 5).
4. Dates collected in Julian days were converted to calendar days.
5. Time of the continuous records was adjusted to Eastern Standard Time.

6. Water levels were checked against manual measurements to check for possible shifts or corrections in the location of the transducer or drift in the data.
7. After the data were processed, hydrographs were constructed and assessed for possible errors. Obvious data errors were removed from the hydrographs; however, the original recorded depth to water was annotated to be erroneous and maintained within the data tables.

Continuous water-level measurements were downloaded every 4 to 6 weeks. Periodic calibration measurements were similar to the calibrations made at the time of installation; however, a small depth correction had to be made to the transducer set depth. Most often, changes in the calibration had to be made whenever the transducer was removed and replaced in the borehole. These adjustments were typically 0.1 ft or less. Performance and results are summarized for each borehole in appendix 1. Electronic drift in the transducers was assumed to be incremental between check measurements and calibration. Hence the data were adjusted linearly, when necessary.

Assessment of Cross-Connections Between Boreholes

Short-term discrete-interval pumping tests were conducted to assess the connections between discrete intervals in different boreholes and to assess impacts on nearby piezometers completed in overburden and shallow bedrock. These tests were conducted on July 11 and 12, August 21, and during September 15-19, 2002. Boreholes were pumped from discrete zones for approximately ¾-hr intervals, while water levels were monitored in adjacent boreholes and piezometers. Continuous data loggers were temporarily installed in wells MW203R and MW204R. Water levels also were monitored using downhole transducers and data loggers in all nearby open shallow boreholes in overburden and bedrock including MW123SR, MW303SR, and MW104 (John Kastrinos, Haley and Aldrich, Inc., written commun., 2002). Data for these cross-connection tests were collected at 1-min intervals. Table 4 shows the location of the pumped intervals; the date, time, and rate of pumping; and which wells were monitored for head.

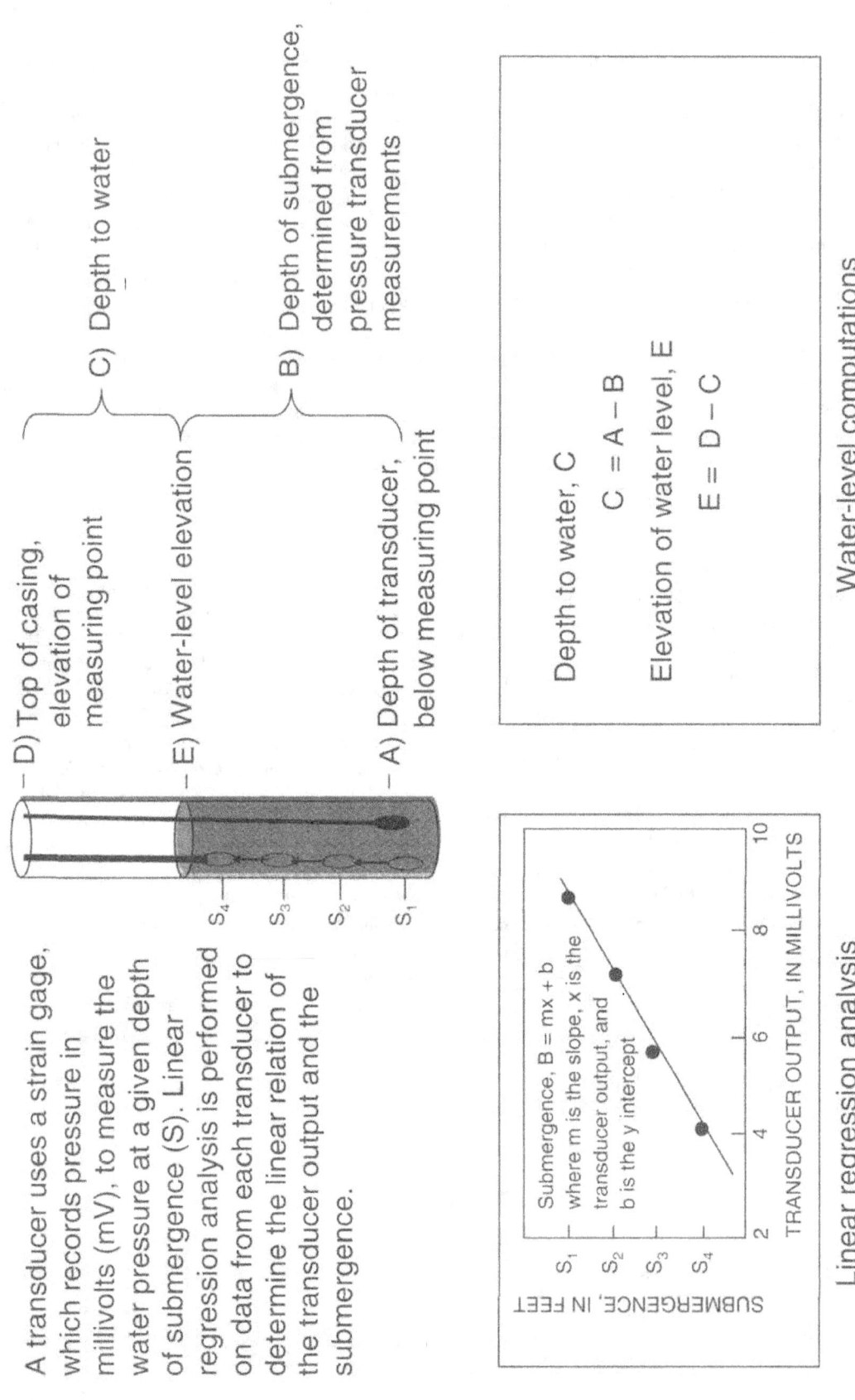

Figure 5. Determination of water levels from transducer data.

Table 4. Data collection information for borehole-pumping cross-connection tests at the UConn landfill study area, Storrs, Connecticut.

[Method of pumping: P, peristaltic pump; S, submersible pump; F, foot valve pump. Boreholes monitored during pumping: D, zones in the boreholes were monitored with Druck PDCR35 transducers; L, open-hole water levels were monitored with Solinst Canada Level Logger downhole data logger. --, no data]

	Pumping zones					
	MW104R Zone 5	MW104R Zone 1	MW122R Zone 2	MW122R Zone 1	MW203R Zone 1	MW104R Zone 5
Method of pumping	P	S	P	S	S	F
Pumping Rate 1						
Date	7/10/02	7/11/02	7/11/02	7/11/02	7/11/02	8/21/02
Start time	19:03	11:14	13:49	15:03	16:58	--
End time	19:34	11:51	14:37	15:33	17:58	--
Rate of pumping, in gallons per minute	0.01	0.25	0.06-0.10	0.20	0.10	Variable
Pumping Rate 2						
Date	7/10/02	7/11/02	7/11/02	7/11/02	7/11/02	--
Start time	19:35	11:55	14:37	15:33	17:38	--
End time	19:45	12:15	14:40	15:46	17:58	--
Rate of pumping, in gallons per minute	0.20	1.00	0.30	0.25	0.50	--
Boreholes monitored during pumping						
MW103R	D	D	D	D	D	D
MW104R	D	D	D	D	D	D
MW105R	D	D	D	D	D	D
MW122R	D	D	D	D	D	D
MW203R	D	D	D	D	D	D
MW204R	D	D	D	D	D	D
MW104 (piezometer)	L	L	L	L	L	L
MW123SR	L	L	L	L	L	L
MW303SR	L	L	L	L	L	L

Description of Precipitation Data

The water levels collected for this investigation were compared to daily precipitation data collected by the National Oceanic and Atmospheric Administration (NOAA) as part of a totally automated observation system used for operational forecasting. The precipitation data are not "official," which means there may be gaps in the record, and there may be errors that are not corrected. The precipitation data were downloaded monthly from the NOAA website for Willimantic, Connecticut (National Oceanic and Atmospheric Administration, 2002). Precipitation, including rain and snow, was reported daily, in hundredths of inches. The daily precipitation uses the period 1 a.m. to 1 a.m. Eastern Standard Time (2 a.m. to 2 a.m. Eastern Daylight Time). Trace precipitation, which was simply identified as T in the NOAA database, was converted to a fixed numeric value of 0.001 inch in the database used to generate precipitation plots.

The daily precipitation measurements are plotted at the same scale as the hydrographs and are shown below the hydrographs. Figure 6A shows the average monthly precipitation collected during each year of the investigation. Average monthly values at the Willimantic gage show a great deal of fluctuation from month to month, so it is difficult to identify periods of drought. For comparison, the 30-year average monthly precipitation values are shown for gaging stations in Bridgeport and Hartford, Connecticut (fig. 6A). The monthly values in the 30-year average range from 3 to 4 in. over the entire year, and there are no clear seasonal-low or seasonal-high values in the long-term averages.

Figure 6B shows the monthly precipitation data in a cumulative plot, where the monthly values are summed over the year. In this form (as in fig. 6A) it is apparent that 1999 had high precipitation in January and in September, and average or below-average precipitation through the rest of the year. The total precipitation for 1999 was similar to the long-term cumulative averages. Figure 6A shows that there are some months with very low precipitation relative to the long-term average monthly values, but their occurrence appears to be sporadic rather than cyclic, and it is difficult to identify a long-term period of drought. The cumulative plots for 2000, 2001, and 2002 show total annual precipitation that is slightly lower than the long-term cumulative averages.

Results of Investigations

In this section, the DZM completion and water-level data are summarized for each borehole. Hydrographs of the manually collected water levels are provided along with the packer inflation and daily precipitation for the same period of record. The hydraulic heads from discrete zones in the DZM system were compared to (1) the heads in adjacent monitoring zones in the same borehole, (2) precipitation data, (3) packer inflation, and (4) the heads measured in nearby DZM systems. Manual water-level measurements and packer inflations for each borehole are shown in data tables in appendix 2. The design, installation, and results of continuous water-level monitoring are summarized for the boreholes that were monitored continuously. Hydrographs showing the manual measurements and the continuous records are shown in appendix 3.

Fluctuations in Water-Level Hydrographs

Fluctuations in water-level data plotted in the hydrographs can be caused by several factors. Hydraulic heads vary in response to recharge, pumping, evapotranspiration, barometric pressure, and tidal forces. Long-term trends in the data and seasonal fluctuations can be seen in the study-area hydrographs constructed from periodic manual measurements. High-frequency fluctuations cannot be seen in the periodic data, but can be seen in the 15-min-interval continuous record. All hydrographs generated in this study were inspected for trends, fluctuations, abrupt changes, and outliers.

Precipitation is the dominant cause of water-level fluctuations. To aid in the interpretation of water-level fluctuations, daily precipitation was plotted along with the hydrographs. The water-level response to precipitation can be immediate or delayed, dramatic or subdued. Long-term cyclic patterns associated with seasonal changes in precipitation can be seen in the hydrographs. Typically during the summer, the combined effect of seasonal increases in evapotranspiration during the growing season and decreases in precipitation causes declines in water levels. During the spring and winter months, the water levels increase in response to the increase in precipitation and the decrease in evapotranspiration.

The continuous water-level data were analyzed to identify possible effects of pumping. These short-term fluctuations typically are cyclic and tend to coincide with regular pumping schedules, which can be weekly or daily cycles. No effects of pumping were identified in the continuous water-level records.

Changes in barometric pressure, which are associated with weather fronts, can cause short-term fluctuations. Even if the transducers are vented to the atmosphere, they still are affected by barometric changes on the aquifer. Typically, there is a lag between the change in barometric pressure and the change in water level; this lag is associated with the barometric efficiency of an aquifer (Todd, 1980). An increase in barometric pressure can cause a decline in the water level.

Earth tides, caused by the moon and sun's gravitational pull, are another cause of short-term fluctuations. Because there are five earth tides with periods of 12 to 24 h, the combined effect is a characteristic semi-diurnal fluctuation in the water level. Some characteristics of earth-tide influences include (1) the cycle occurs 50 min later each day; (2) the daily troughs of the water-level declines coincide with the daily transit of the moon, and are nearly coincident with the time that gravitational forces are at a maximum; (3) the largest fluctuations in the water levels coincide with the full moon and new moon (syzygy, when the sun, moon and earth are in alignment); and (4) small amplitude changes in the water levels occur when the earth tide forces are at a minimum (Todd, 1980). This fluctuation typically impacts semi-confined and confined aquifers. The continuous water-level data in this investigation show fairly strong effects of tidal fluctuations. The uppermost zones in the DZM systems show lower amplitude responses to the earth tides. Gravitational forces were computed for November 2001 for latitude 41° 48' 45" and longitude 72° 16' 15" at an elevation of 600 ft above NAVD 88 (fig. 7) (Harrison, 1971).

(A)

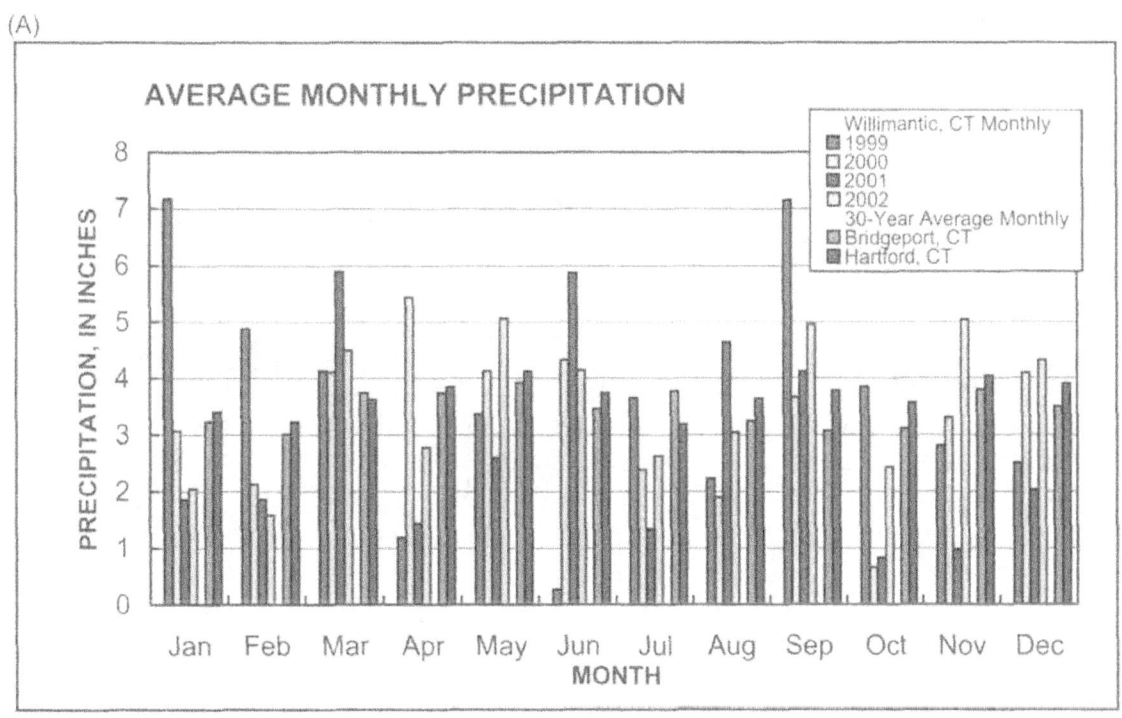

(B)

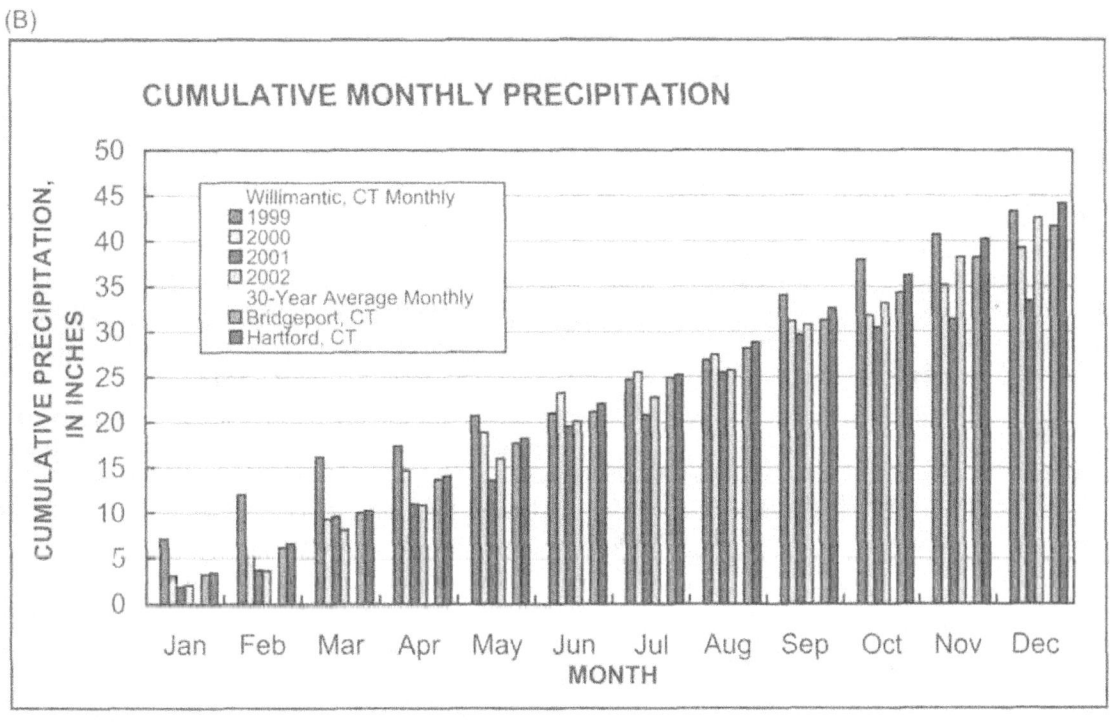

Figure 6. (A) Average monthly precipitation at three locations in Connecticut. (B) Cumulative monthly precipitation at three locations in Connecticut.

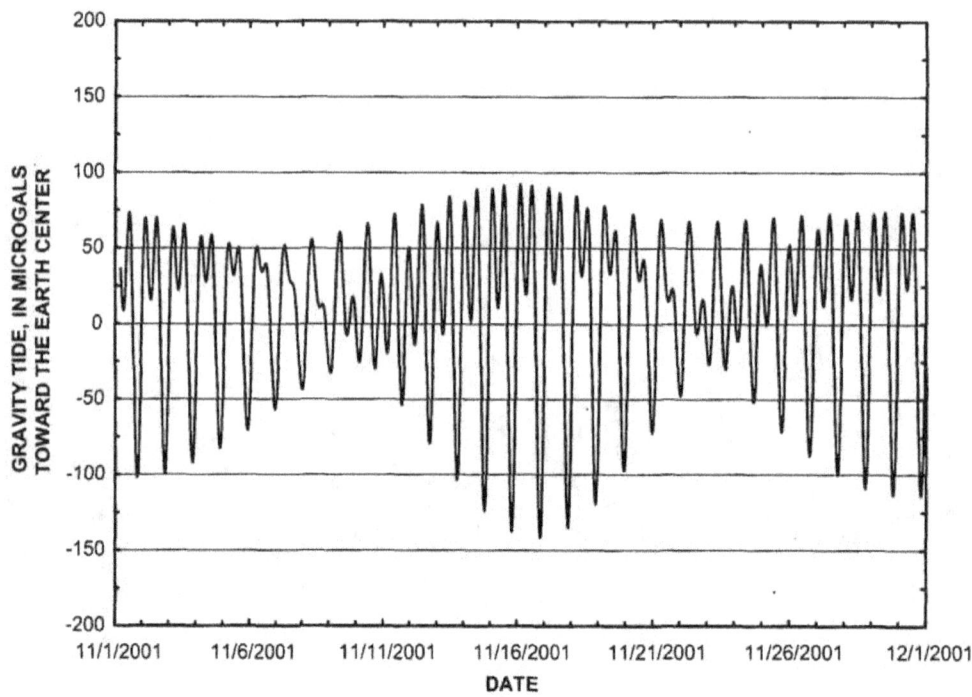

Figure 7. Calculated gravimetric forces for latitude 41° 48' 45", longitude 72° 16' 15" at an elevation of 600 feet for the period between November 1, 2001 and December 1, 2002.

Problems Common to Discrete-Zone Monitoring Systems and Data Loggers

Problems common to many of the DZM systems and leading to periods of missing or unreliable data include failing packers, leaking pressure lines, freezing water lines, failing pressure gages used to monitor the packer inflations, and overflowing water in individual CMT channels. Other problems unique to the continuous-monitoring systems also can lead to the loss of continuous record. These include moisture problems in the data-logger shelter, shifted or damaged transducers, and battery or data-logger failure.

Packer failure and leaking pressure systems. Immediately after the DZM systems were installed, packer inflation was monitored every few days until it stabilized. Sometimes fittings at the top of the borehole had to be replaced. The DZM systems in two of the boreholes, MW101R and MW103R, had to be removed, and leaking packers had to be replaced. The packers maintained inflation fairly well, unless water was pumped from the discrete zones. Consequently, packer inflations were checked and maintained after all water-quality sampling events. In general, packer inflations were "topped off" about every 4 to 8 weeks or as needed.

Typically the packers were filled to a total pressure of about 20 psi over ambient conditions. During field testing, hydraulic head differences in some boreholes were maintained with packer inflations as low as 6 psi. In cases where packer pressure fell below 10 psi, the data points were reviewed, and if the water levels in adjacent zones converged, then the head measurements were identified as "packer failure," and these data were not included in the hydrographs.

Freezing conditions. The inflation lines, which are filled with water, were subject to freezing during winter months. During these periods of frozen lines, the packer inflations could not be determined. The packer and pressure integrity, however, was considered to be sufficient if differential heads were maintained. The water-level records indicate that if the lines remained frozen, the differential heads were maintained. When the lines thawed, however, the inflation lines, gages and packers would leak, causing the convergence of water levels. After the winter of 2001, all of the inline gages had to be replaced or removed because of damage caused by freezing.

The frozen inflation lines were a problem typically when the packers had to be reinflated after sampling during the winter months. Multiple steps were taken to attempt to prevent the inflation lines from freezing including:

1. The shelters and the casing were wrapped with insulation to retain heat in the borehole.

2. The inflation lines were lowered into the borehole and below the frost line, and a string was attached to the lines to aid in their retrieval. This method may have temporarily delayed the inflation lines from freezing, but did not prevent it.

3. Injecting air bubbles into the top of the inflation lines was attempted on selected boreholes, but this action did not prevent freezing.

4. The addition of a light non-aqueous phase food additive (with a very low freezing point) to the top of the inflation lines was considered. Because this oil floats, opening the top of the inflation line would force the oil out of the line where it could be captured and reused the following year. However, this procedure was not used because of concerns of the potential for leakage or spillage.

5. Another alternative, which was considered, was to put a solar-powered active heating source into the well shelters. Because of concerns with vandalism, however, this method was not attempted.

During the winter months, access to some of the channels in the DZM systems also were periodically and temporarily blocked because of freezing or because of friction in the channel, which prevented the water-level meter from reaching the water surface. Hence, there are periods of missing measurements in water levels and pressure readings.

Flowing or dry conditions. Only one borehole in the study area, MW201R, flows for part of the year. During that period of record, the head is known to be higher than the top of casing. During flowing conditions, the heads on the hydrograph plot at the height of the top of the casing, and the associated data points are labeled as "flowing." A similar problem occurs in the overburden piezometers, MW104 and MW109, when the water level declines below the bottom of the well. The affected portions of the records are labeled as "dry."

Moisture conditions. High moisture levels in the data-logger shelter adversely affected the DZM systems that were monitored with continuous water-level monitoring equipment. Desiccants were used in the data-logger shelter to keep the transducers as dry as possible. Because the transducers used in the DZM systems are vented transducers, they are subject to blockage by moisture. Adversely affected data were omitted from the record.

Shifted or damaged transducers. Occasionally, transducers were moved during sampling events, and they were not returned to the same place. This problem was identified in the calibration process. One transducer or cable in the DZM system in MW101R was damaged and replaced. The adversely affected data were removed from the hydrograph.

Battery or data-logger failure. If the battery power dropped below the operating voltage, the data logger stopped recording. Data that had already been saved on the CR10X data loggers were not affected. All periods of missing data or data that were corrupted are shown as missing data on the continuous water-level hydrographs.

Results for Boreholes at the UConn Landfill Study Area

This section discusses data collection and compilation for each borehole where discrete-zone hydraulic measurements were made in the UConn landfill study area. As applicable, the discussion for each borehole includes (1) manual open-hole water-level measurements, (2) manual DZM system water-level measurements, (3) continuous DZM systems water-level measurements, and (or) (4) cross-connection pumping tests and connectivity analyses. The DZM completion for MW101R is shown in figure 4 and the DZM completions for the other boreholes in this study are shown in figure 8.

ELEVATION, IN FEET ABOVE NORTH
AMERICAN VERTICAL DATUM OF 1988

ELEVATION, IN FEET ABOVE NORTH
AMERICAN VERTICAL DATUM OF 1988

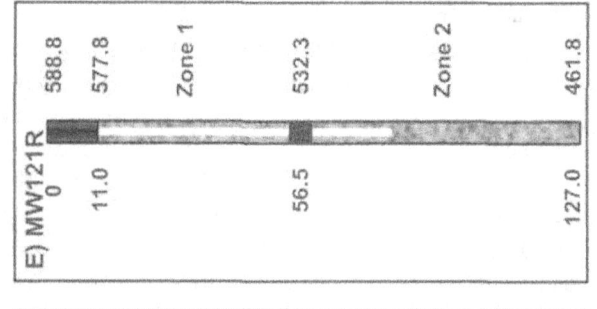

E) MW121R

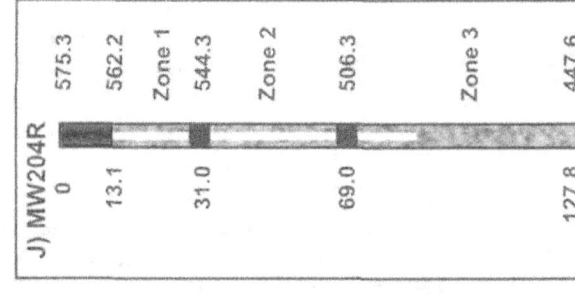

J) MW204R

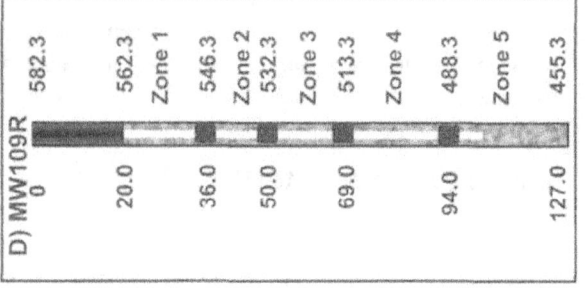

D) MW109R

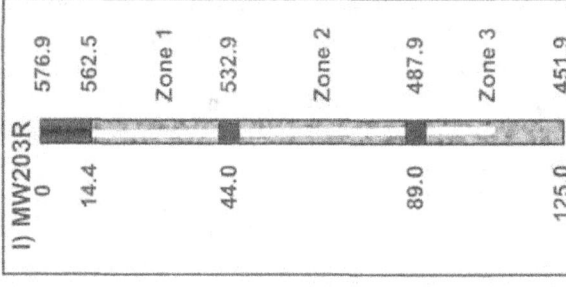

I) MW203R

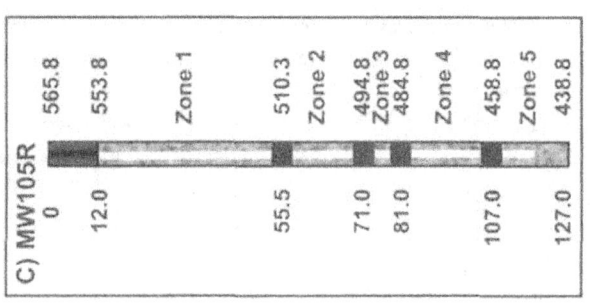

C) MW105R

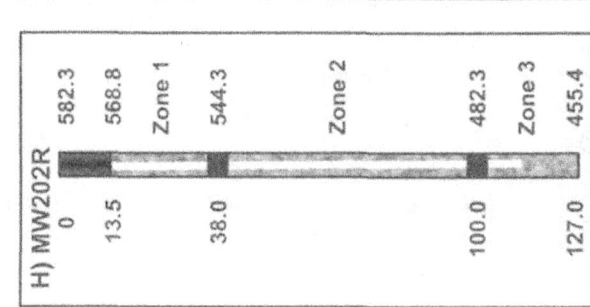

H) MW202R

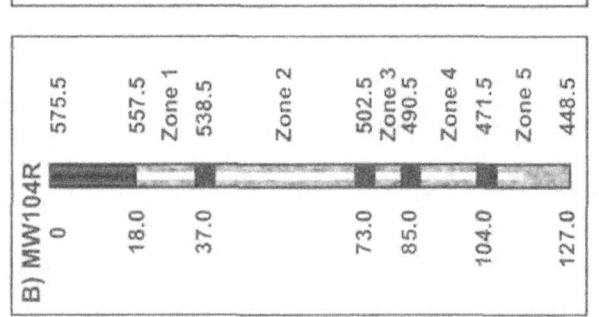

B) MW104R

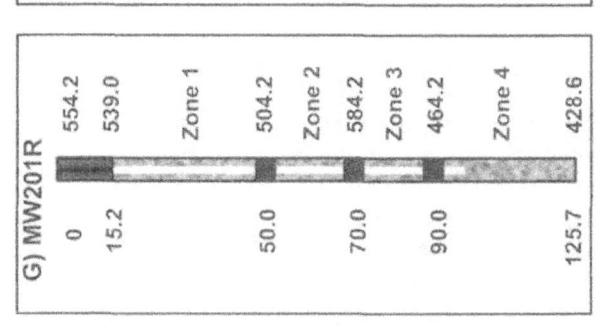

G) MW201R

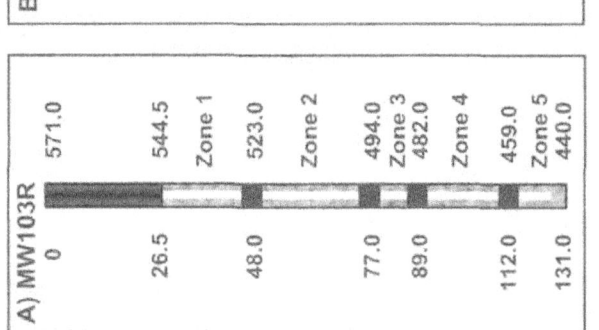

A) MW103R

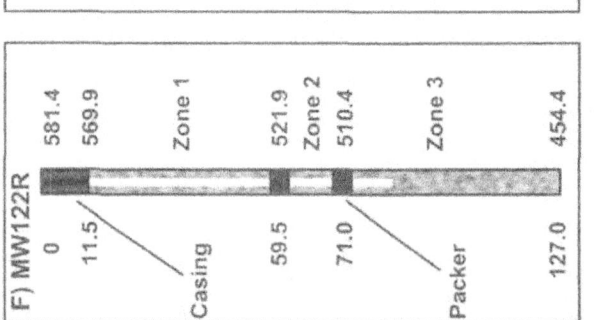

F) MW122R

DEPTH, IN FEET BELOW TOP OF CASING

DEPTH, IN FEET BELOW TOP OF CASING

Figure 8. Discrete-zone monitoring system completions for boreholes in the UConn landfill study area, Storrs, Connecticut.

Borehole MW101R

Borehole MW101R is located on the northern flank of the landfill near a wetland (fig. 1). The borehole was drilled using air-hammer rotary methods and was completed as an open hole with 23 ft of 6-in-diameter steel casing and open to a depth of about 126 ft below land surface. Below the casing, the DZM system isolates four zones at elevations of 530.5-514.5 ft; 512.5-493.5 ft; 491.5-485.0 ft; and 483.0-426.5 ft above NAVD 88 (fig. 4). A DZM system was successfully installed on June 14, 2000. Attempts were made to install the packer system on May 31, June 8, and June 9, 2000, but for various reasons these packer installations were not successful. Transmissive fractures were identified in Zones 1, 2, and 3 at depths of 35, 45, and 65 ft below the top of the casing (Johnson and others, 2002). Three packers were installed in MW101R at center depths of 40.0, 61.0, and 69.5 ft below the top of casing (fig. 4). Four ports were installed at depths of 35.0, 46.0, 65.0, and 72.0 ft below the top of casing.

A hydrograph was constructed with manual measurements for borehole MW101R (fig. 9). The lowest zone in the borehole has the highest head, and the uppermost zone has the lowest head. For example on March 14, 2001, the head in Zone 4 was 552.57 ft above NAVD 88, higher than the heads in Zones 1, 2, and 3, which were 551.05, 551.20, and 552.26 ft above NAVD 88, respectively. Thus the driving potential between Zone 4 and Zone 3 is upward. Similarly, the head in Zone 3 is higher than the head in Zone 2, which is higher than the head in Zone 1. This pattern is consistent over the entire period of record.

The hydrograph for MW101R shows an upward driving potential between all discrete zones in the borehole. The head in a nearby piezometer, MW101, which is screened in the overburden, is lower than any of the heads in the bedrock well (Haley and Aldrich, Inc. and others, 2000, table 5). These results indicate an upward hydraulic gradient in the bedrock and between the bedrock and overburden.

All zones respond to precipitation as seen by the increased heads associated with the rainfall on September 15, 2000, December 16, 2000, June 17, 2001, and May 13, 2002. The overall water levels, however, did not decline much during the period of drought from June 2001 to September 2001, showing only a few feet of decline.

The head and transmissivity values estimated with two heat-pulse flowmeter profiles are shown in table 5. The hydrau-lic heads are consistent with the DZM system results. On August 26, 1999, a heat-pulse flowmeter measured upward flow of 0.018 gal/min between the fractures at 65 and 35 ft (Johnson and others, 2002). Although the magnitude of head difference changed over the period of record, the driving potential was consistently upward between zones in the borehole. This implies that if multiple heat-pulse flowmeter measurements were taken over the same period of record, the rate of vertical flow would change, but the direction of measurable flow would always be upward. In addition, the DZM system provided information about the magnitude of differential heads between fractures that did not have flow measurable at the resolution of the heat-pulse flowmeter.

The magnitudes of the differences in head between zones simulated with Paillet's program (2000) are very similar to the differences between Zone 1, Zone 2, and Zone 3 measured with the DZM system. The elevations of the water levels are lower than the water levels collected during August 2001 and August 2002. The water level for Zone 2, however, which is the most transmissive zone in the borehole, is similar to the open-hole heads obtained in August and September 1999, when the flowmeter logs were collected. The simulated ambient and pumped flow profiles are correlated, and the model SS error was 0.0006 gallons per minute, squared $(gal/min)^2$.

The transmissivities obtained with the heat-pulse flowmeter logging and modeling were greater than the transmissivity estimates determined with discrete-interval packer testing (Haley and Aldrich, Inc. and others, 2002). Discrete-interval constant-rate withdrawal tests gave an estimated transmissivity of 3.1 ft^2/d for the fracture at a depth of about 45 ft. Using two different methods of interpretation, the transmissivity of the lower zone at 65 ft was estimated at 3.1 ft^2/d and 18 ft^2/d (Haley and Aldrich, Inc. and others, 2002). Results of the modeled flowmeter logging are between these two estimates.

The upward driving potential in MW101R and between MW101R and piezometer MW101 and the proximity of MW101R and MW101 to a perennial wetland is consistent with the interpretation that MW101R and MW101 are located in a ground-water discharge area. The heads in MW101R are consistently lower than the heads in MW103R and in the other boreholes near the former chemical-waste disposal pits, indicating that MW101R is in a ground-water discharge location.

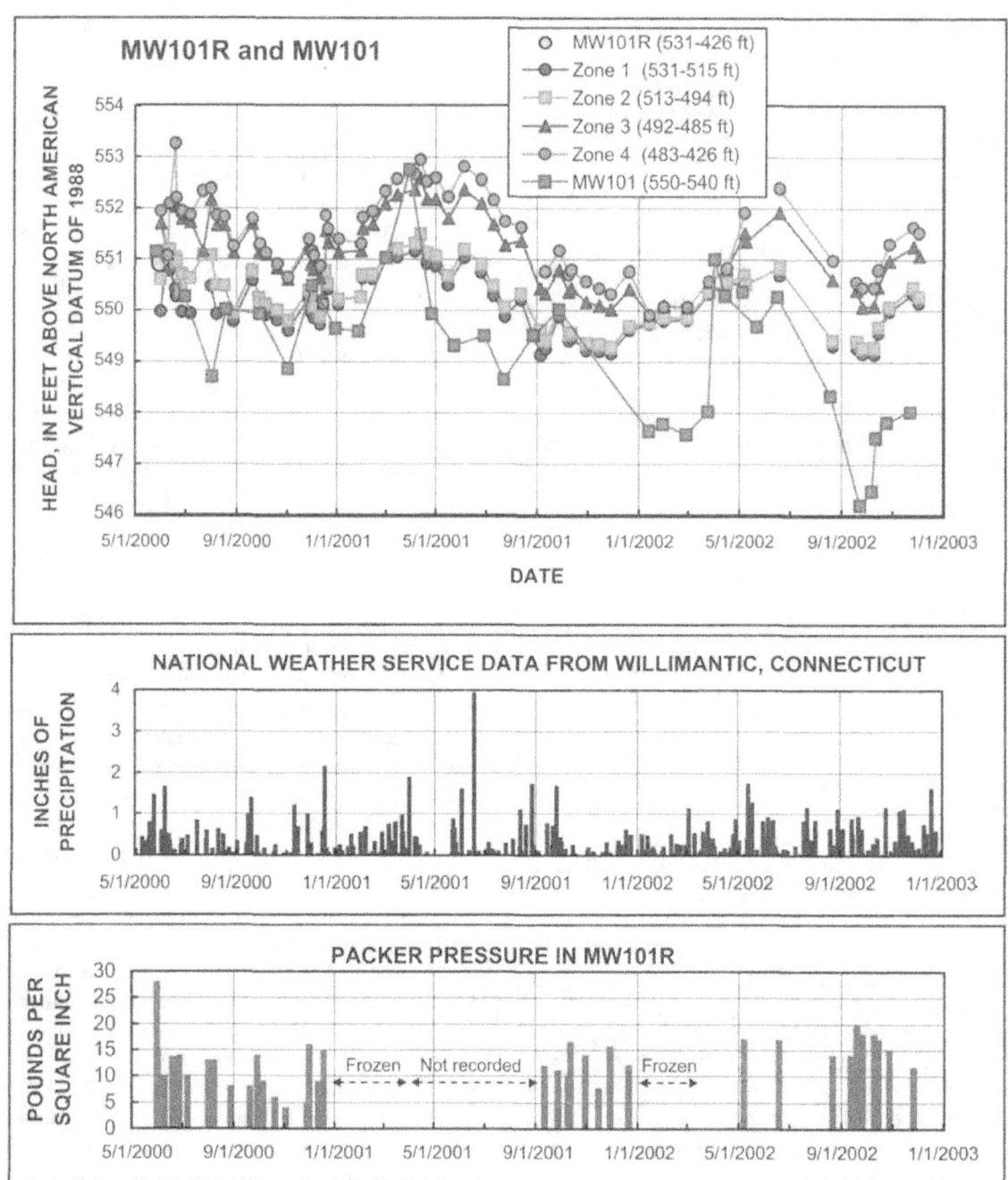

Figure 9. Hydrographs for MW101R and MW101 in the UConn landfill study area, Storrs, Connecticut. Values in parentheses next to the symbols indicate the elevation of open hole for that monitoring zone. Precipitation and packer-pressure data are shown below the hydrograph. Gaps in the record indicate missing or erroneous data that were omitted.

Table 5. Heads and transmissivity derived from heat-pulse flowmeter measurements in MW101R in the UConn landfill study area, Storrs, Connecticut, August 30, 1999.

[DZM, discrete-zone monitoring (system). Elevation is based on NAVD 88]

Depth, in feet below top of casing	Transmissive fracture in DZM zone	[1]Relative head, in feet	Water-level elevation, in feet	Transmissivity, in feet squared per day	Proportion of transmissivity, in percent
35.0	Zone 1	-1.67	547.54	2.1	8
45.0	Zone 2	-0.32	548.89	14.3	54
65.0	Zone 3	0	549.21	10.1	38

[1]Relative head is the magnitude of head for the zone relative to the head of the lowermost zone.

Transducers were installed in Zones 1, 2, and 3 at MW101R. Over much of the record, the continuous record showed a fairly poor correlation with the measured values. A plot of manual measurements and corrected continuous water-level measurements data are shown in appendix 3A. A table shows the calibration values and transducer set depths that were used to compute the water levels (appendix 1A). Large correction values of up to 1.5 to 2 ft in the set depths were needed after the transducers were temporarily removed from the borehole for ground-water sampling. Because the water levels were near or above land surface over most of the year, the data logger and the vented ends of the transducers were subject to moisture, which may have contributed to poor correlation with manually measured values. In mid-June 2002, the transducer in Zone 3 appears to have had the vent line blocked, which could account for the nearly 2-ft error in the water level measured by the transducer as compared to the manual measurements. This transducer was replaced in September 2002. Once the transducer data were corrected to the manual measurements, the hydrographs identified more fluctuation in water levels than were recorded in the manual measurements. All transducers showed water-level increases associated with precipitation events. None of the transducers showed strong responses to the gravimetric forces, but Zone 3 showed the strongest response of any of the zones under full- and new-moon conditions.

Borehole MW103R

Borehole MW103R is located at the base of the slope of the landfill on the northwestern side of the landfill (fig. 1). Borehole MW103R was drilled using air-hammer rotary methods, was completed with 26.5 ft of 6-in-diameter casing, and was open below the casing to 129 ft below land surface. Four packers were installed during September 2000 at center depths of 48.0, 77.0, 89.0, and 112.0 ft below top of casing (fig. 8A). Five ports were installed at depths of 32.0, 54.0, 82.0, 100.0, and 119.0 ft below top of casing. Below the steel casing, the DZM system isolates five zones at elevations of 544.5-524.0; 522.0-495.0; 493.0-483.0; 481.0-460.0; and 458.0-440.0 ft above NAVD 88. Because the original packers did not maintain sufficient inflation, they were replaced in November 2000. Two transmissive fractures were identified in Zone 1 at about 30 and 33 ft below the top of casing, and another transmissive fracture was identified in Zone 3 at 82.5 ft below the top of casing (Johnson and others, 2002). Transmissive fractures identified in Zones 1 and 3 were targeted for water-quality sampling.

A hydrograph of MW103R shows that the water levels in the lower four zones are fairly similar and have a higher hydraulic head than the uppermost zone in the borehole (fig. 10). Over most of the period of record, the water levels in Zones 2, 3, 4, and 5 are similar to and higher than the head in Zone 1. These results indicate an upward driving potential, consistent with the interpretation that MW103R is located in a discharge zone. The heads measured in the DZM system are not in agreement with the flowmeter data collected in 1999 and described below.

The hydrograph (fig. 10) indicates that all zones show a response to precipitation. During a period of minimal precipitation, from spring of 2001 through December 2001, heads from all zones showed about a 9-ft decline. The heads in all zones during this period of drought are very similar, showing less of a driving potential than during times with more precipitation.

Heat-pulse flowmeter measurements collected under ambient conditions on September 17, 1999, indicated that ground water entered borehole MW103R at about 32 ft below the top of casing and flowed downward, where it exited the borehole at 82 ft below the top of casing. Under pumping conditions, most of the water entered the borehole from the fractures near a depth of 30 ft, no outflow occurred at 82 ft, and a minor amount (less than 5 percent) of water entered at 118 ft below the top of casing. The system was simplified and modeled as a two-fracture system. The head and transmissivity computed from the two heat-pulse flowmeter profiles are shown in table 6. The model produced an SS error of less than 0.0001 $(gal/min)^2$. The transmissivity results confirm that the upper zone is the dominant transmissive fracture in the borehole. The modeled heads show a downward driving potential, which is contradictory to what was observed in the long-term head records in the DZM system. This is the only borehole in the study where the heads from the DZM system differed with the results of the heat-pulse flowmeter. One possible explanation is that the heat-pulse flowmeter results were collected during an extreme period of drought when the direction of driving potentials may have been temporarily reversed. The transmissivities obtained with the heat-pulse flowmeter logging and modeling are similar to the results of discrete-interval packer testing, which identified the upper zone to have a transmissivity of 19 ft^2/d and the lower zone to have a transmissivity of 6.6 ft^2/d (Haley and Aldrich, Inc. and others, 2002).

The heads were compared to nearby piezometer, MW103, completed in the overburden, and to borehole MW101R farther to the north. The hydrograph indicates a consistent upward driving potential between all discrete zones in borehole MW103R and piezometer MW103 (fig. 10). The heads in all zones in MW103R are consistently higher than the elevation of heads in MW101R. Borehole MW103R is considered to be at an intermediate location along the north-south flowpath extending from the ground-water divide towards the wetland to the north of the landfill.

Transducers were installed in Zones 1, 2, 3, and 5. The continuous record showed fairly good correlation with the measured values; these results are presented in appendix 3B. The continuous record shows more fluctuation in water levels than were recorded in the manual measurements.

Only minor adjustments were applied to the calculation of water levels from transducer data. The continuous record for Zone 1 showed the most variation in water-level response to precipitation events. The heads in Zones 2, 3, and 5 also showed a response to precipitation, although the water-level response was more subdued than in Zone 1.

Hydraulic heads in Zones 2, 3, and 5 showed response to the gravimetric forces and a well-defined semi-diurnal pattern. Under full- and new-moon conditions, the amplitude of the daily water-level changes was about 0.08 ft in Zone 2; 0.12 ft in Zone 3; and about 0.06 ft in Zone 5. The uppermost zone, Zone 1, does not show gravimetric tidal fluctuation, indicating it is unconfined.

Table 6. Heads and transmissivity derived from heat-pulse flowmeter measurements in MW103R in the UConn landfill study area, Storrs, Connecticut, September 17, 1999.

[DZM, discrete-zone monitoring (system). Elevation is based on NAVD 88]

Depth, in feet below top of casing	Transmissive fracture in DZM zone	[1]Relative head, in feet	Water-level elevation, in feet	Transmissivity, in feet squared per day	Proportion of transmissivity, in percent
32.0	Zone 1	0.68	562.75	21.0	75
82.0	Zone 3	0	562.07	6.9	25

[1] Relative head is the magnitude of head for the zone relative to the head of the lowermost zone.

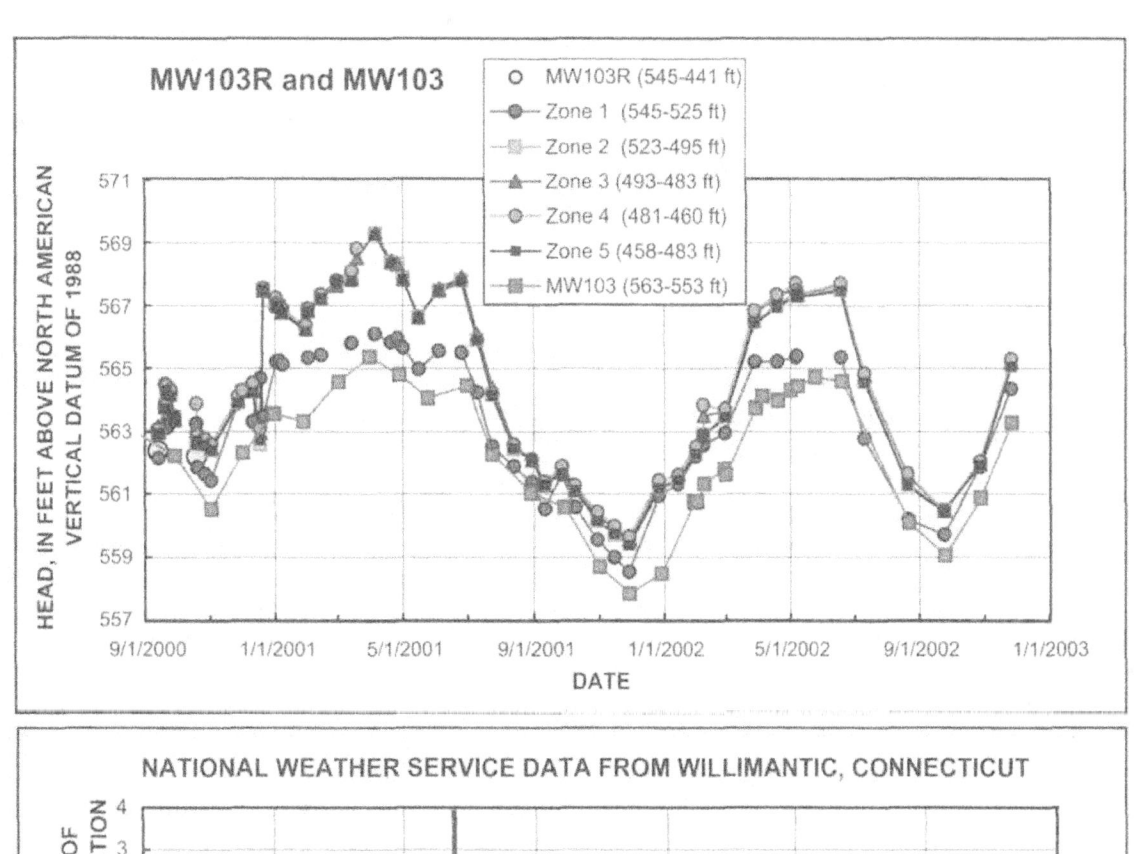

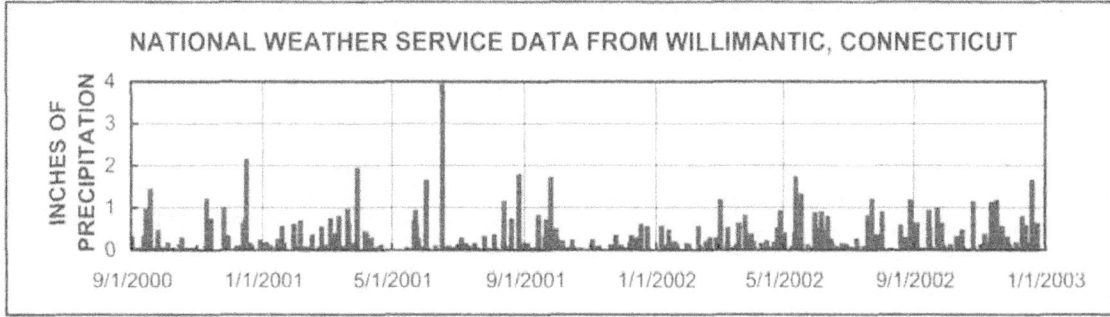

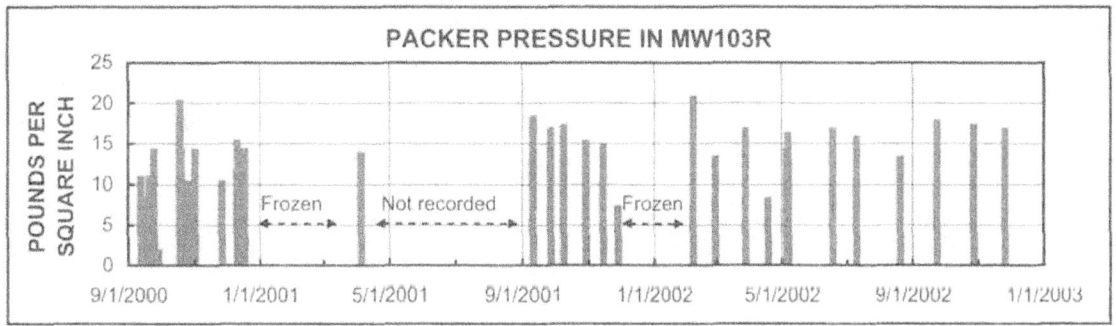

Figure 10. Hydrographs for MW103R and MW103 in the UConn landfill study area, Storrs, Connecticut. Values in parentheses next to the symbols indicate the elevation of open hole for that monitoring zone. Precipitation and packer-pressure data are shown below the hydrograph. Gaps in the record indicate missing or erroneous data that were omitted.

Borehole MW104R

Borehole MW104R was drilled using air-hammer rotary methods and was completed with 18 ft of 6-in-diameter casing and open below the casing to about 125 ft below land surface. A total of four packers was installed at center depths of 37.0, 73.0, 85.0, and 104.0 ft below the top of casing (fig. 8B). Five ports were installed at depths of 24.0, 45.0, 81.0, 88.0, and 110.0 ft below the top of casing. Below the steel casing, the DZM system isolates five zones at elevations of 557.5-539.5; 537.5-503.5; 501.5-491.5; 489.5-472.5; and 470.5-448.5 ft above NAVD 88. Transmissive fractures identified in Zones 1 and 5 were targeted for hydraulic testing and water-quality sampling (Johnson and others, 2002). The open-hole head, collected prior to the installation of the DZM system on September 18, 2000, was dominated by the head in Zone 5, indicating this zone is the most transmissive zone in the borehole. Previous investigations, however, indicate that the upper and lower transmissive zones have nearly the same transmissivity (Johnson and others, 2002).

All zones show fairly sharp increases in head in response to precipitation and gradual declines in head in the absence of precipitation (fig. 11). In each zone in MW104R, the water levels show a decline of about 10 ft from the spring of 2001 to the winter of 2001 in response to seasonal variation and drought.

Heat-pulse flowmeter measurements collected on August 30, 1999, in MW104R indicate that ground water entered near the top of the borehole at a depth of 24 ft and flowed downward exiting the borehole near the bottom at a depth of 108 ft below the top of casing. Water levels collected in the DZM system are consistent with the flowmeter profiles. Water levels collected from Zones 1-4 show progressively lower heads, indicating downward driving potentials (fig. 11). Zones 4 and 5 have the same head over most of the profile.

The heads computed from heat-pulse flowmeter profiles have a similar downward driving potential, but the magnitude of the difference computed with the flow profiles is much less than what has been measured in the DZM system. The computed head of the lower zone (554.38 ft above NAVD 88) is similar to the DZM heads collected during the summer months in 2000 and 2001, and the computed head of the upper zone (555.03 ft above NAVD 88) is substantially lower than the heads observed in the upper zone of the DZM system during the summer months of 2000 and 2001. This difference may be because the flowmeter results were collected in August 1999, during a time when there was less precipitation and recharge, and hence less downward driving potential.

The transmissivities computed from the two heat-pulse flowmeter profiles are shown in table 7. The model produced an SS error of 0.0001 (gal/min)^2. The results show the two zones to be almost equally transmissive. The transmissivity of the lower zone is somewhat lower than the estimates of transmissivity determined with packer testing methods, which estimated a transmissivity of 22 ft^2/d (Haley and Aldrich, Inc. and others, 2002).

In piezometer MW104, which is completed in the overburden near MW104R, the heads are slightly higher than the heads in Zone 1 in MW104R, indicating a downward driving potential between the overburden and bedrock. Moreover, when the borehole liner was removed from MW104R for geophysical logging and sampling, the head in MW104 dropped to the bottom of the piezometer indicating a very strong connection between the bedrock and the overburden at this location (John Kastrinos, Haley and Aldrich, Inc., written commun., 2000). This observation indicates the importance of a borehole liner or a DZM system to minimize flow created by the driving potential between the overburden and the bedrock.

The heads in the uppermost zone, Zone 1, are consistently higher than the heads in all other zones in MW104R. The heads in the upper zone of MW104R are often at the same elevation as the head in the upper zone of MW204R. The hydraulic head in the upper zone of MW104R is consistently at a higher elevation than all nearby boreholes to the north and south of the former chemical-waste disposal pits, including MW203R, MW103R, MW122R, and MW105R. These data are consistent with the conceptual flow model for the site, which considers this part of the study area to be a local recharge zone near the ground-water divide.

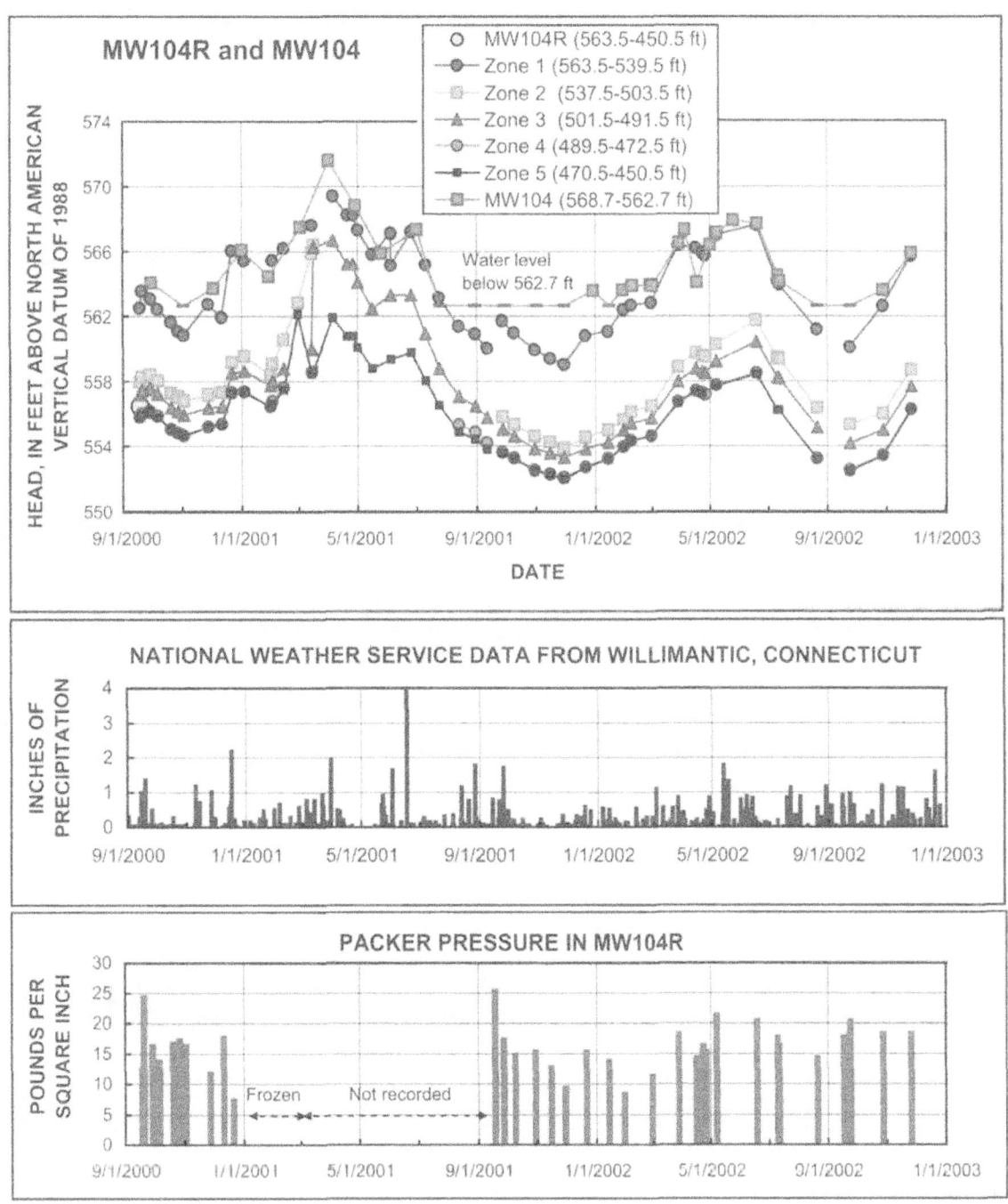

Figure 11. Hydrographs for MW104R and MW104 in the UConn landfill study area, Storrs, Connecticut. Values in parentheses next to the symbols indicate the elevation of open hole for that monitoring zone. Precipitation and packer-pressure data are shown below the hydrograph. Gaps in the record indicate missing or erroneous data that were omitted.

Table 7. Heads and transmissivity derived from heat-pulse flowmeter measurements in MW104R in the UConn landfill study area, Storrs, Connecticut, August 30, 1999.

[DZM, discrete-zone monitoring (system). Elevation is based on NAVD 88]

Depth, in feet below top of casing	Transmissive fracture in DZM zone	[1]Relative head, in feet	Water-level elevation, in feet	Transmissivity, in feet squared per day	Proportion of transmissivity, in percent
24.0	Zone 1	0.65	555.03	15.1	54
108.0	Zone 5	0.0	554.38	12.9	46

[1] Relative head is the magnitude of head for the zone relative to the head of the lowermost zone.

Transducers were installed in Zones 1, 3, and 5. The continuous record showed fairly good correlation with the measured values (appendix 3C). Some minor shifts in the transducer set depths were needed after the transducers were temporarily removed from the borehole for ground-water sampling. The continuous record for Zone 1 showed more variation in the water levels than the deeper Zones 3 and 5. The abrupt increases in water level in Zone 1 are associated with precipitation events. The heads in Zones 3 and 5 also showed a response to precipitation, although the response was more subdued than in Zone 1.

Zones 3 and 5 showed a strong response to the gravimetric forces and a well-defined semi-diurnal pattern. Under full- and new-moon conditions, the amplitude of the daily water-level changes was 0.2 ft in Zone 5 and 0.1 ft in Zone 3, which indicates that Zone 5 is more transmissive than Zone 3. The uppermost zone, Zone 1, does not show gravimetric tidal fluctuation, indicating it is unconfined.

Borehole MW105R

Borehole MW105R is 300 ft southwest of the landfill and approximately 460 ft south of the former chemical-waste disposal pits. MW105R was drilled using air-hammer rotary methods and was completed with 12 ft of casing and open below the casing to about 123 ft below the land surface. A total of four packers was installed in August 2000 at center depths of 55.5, 71.0, 81.0, and 107.0 ft below the top of casing (fig. 8C). Five ports were installed at depths of 50.0, 63.0, 74.0, 90.0, and 110.0 ft below the top of casing. Below the steel casing, the DZM system in MW105R isolates five zones at elevations of 553.8-511.3; 509.3-495.8; 493.8-485.8; 483.8-459.8; and 457.8-438.8 ft above NAVD 88. The open-hole head collected on August 9, 2000, is dominated by the head in the lowermost zone (Zone 5), but is still influenced by the head in Zone 3. This is consistent with the interpretation of heat-pulse flowmeter and hydraulic testing methods that identified the fractures in Zone 3 as transmissive and the fractures in Zone 5 as twice as transmissive as the fractures in Zone 3 (Johnson and others, 2002). Both of these fractures were targeted for water-quality sampling

Over most of the period of record, driving potentials were consistently downward (fig. 12). Some minor variations and reversals in driving potentials occurred in Zones 1 and 2 during summer. From August 2000 to November 2000, the driving potential was upward from Zone 2 to Zone 1. During periods of recharge (February 2001 to April 2001 and February 2002 to June 2002), the head in Zone 1 was higher than the head in Zone 2, and the driving potential was downward. The heads in Zones 1, 2, and 3 are consistently higher than the heads in Zones 4 and 5, indicating an overall downward potential for flow in the borehole.

All zones responded to precipitation. For example, all zones showed minor increases during the precipitation on December 17, 2000; March 30, 2001; and June 17, 2001. The heads within a single zone show as much as 7 ft of decline from spring 2001 through December 2001 in response to seasonal variation and drought.

No ambient flow was measured with the heat-pulse flowmeter on August 30, 1999; however, flow profiles were col-

lected under two different pumping rates and were used to simulate the hydraulic head and transmissivity. For the calculations, the ambient heads were set to zero, because no ambient flow was measured with the heat-pulse flowmeter. In actuality, there may have been differences in head, which caused a minor vertical flow that was less than the resolution of the heat-pulse flowmeter. The hydrograph of the discrete-intervals indicated downward driving potentials in August 2001 and 2002.

Transmissivity values from discrete-interval packer testing were estimated at 2.0 ft^2/d for the fracture at a depth of about 74 ft below top of casing and 13 ft^2/d for the fractures at depths of 111 to 113 ft below the top of casing (Haley and Aldrich, Inc. and others, 2002). The results from the heat-pulse flowmeter measurements (table 8) are higher than the packer-test results, but still within an order of magnitude.

For the entire period of record, the heads in MW105R were consistently lower than the heads in MW104R and MW203R to the north. The heads in MW105R were higher than the heads in MW201R to the south, indicating that MW105R is in an intermediate location along a southern flowpath. The hydrograph and the directions of driving potentials indicate that seasonally, the top part of MW105R discharges upward, but there is also a downward driving potential from the top of the borehole to the bottom of the borehole, consistent with a borehole located in an intermediate setting along a flowpath southwest of the landfill.

Continuous water-level data were collected in four zones in MW105R from November 2001 to November 2002 (appendix 3D). Transducers were installed in Zones 1, 2, 3 and 5. The continuous data show the same seasonal trends as the manual measurements. In addition, the continuous measurements capture minor increases in ground-water levels in response to precipitation events and effects of tidal forces. The continuous water-level record and manual measurements from November 11, 2001, to February 1, 2002, are shown in figure 13. The water levels increase over the period. The manual and continuous records indicate a change in the driving potential between Zone 1 and Zones 2 and 3.

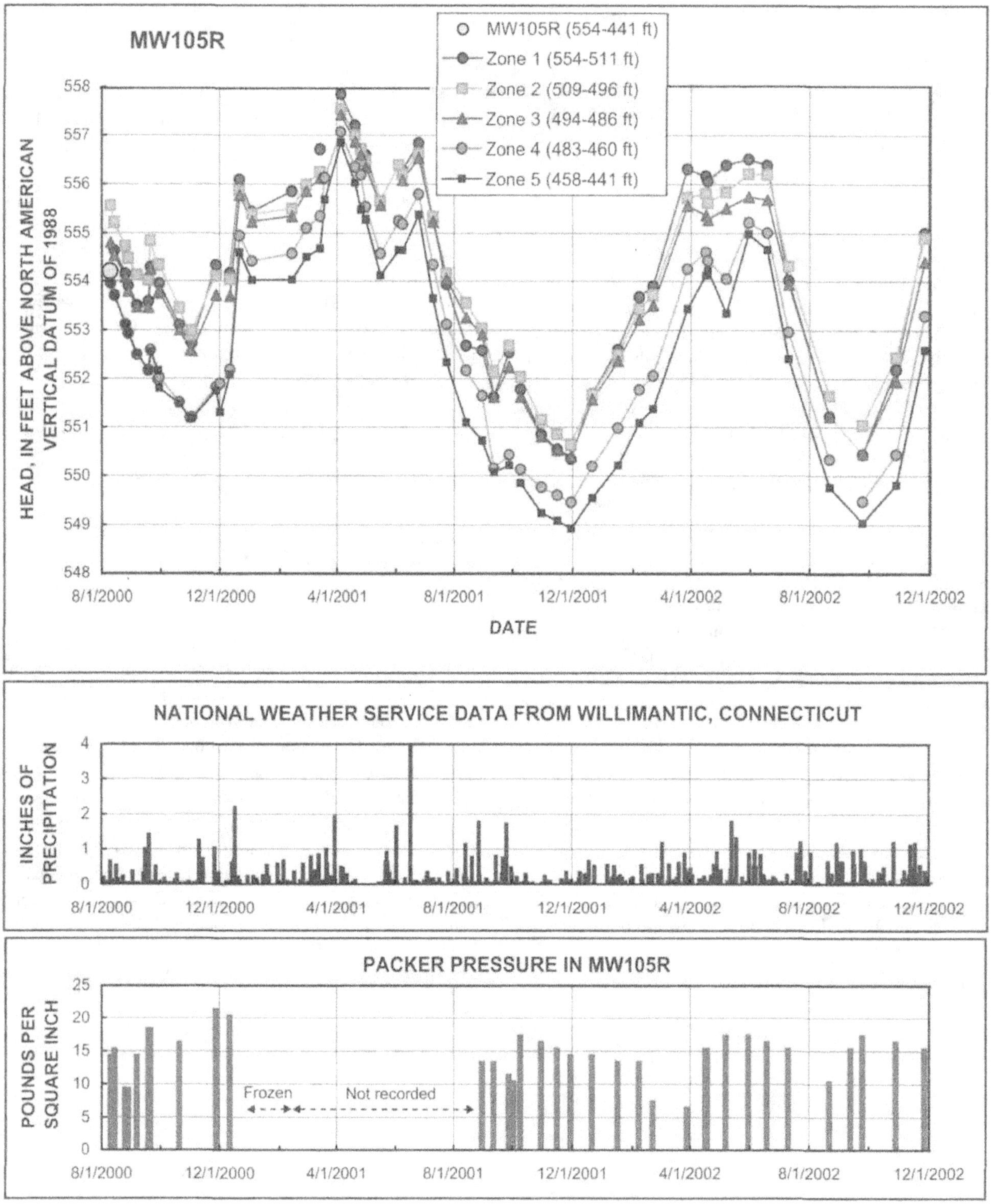

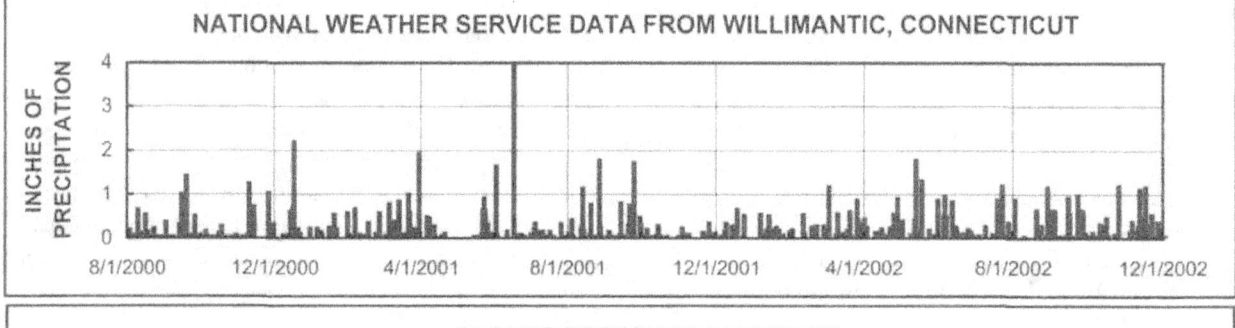

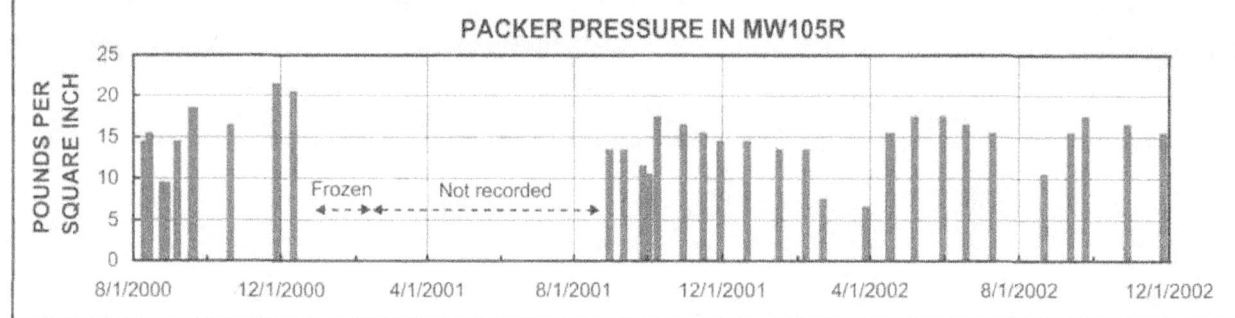

Figure 12. Hydrographs for MW105R in the UConn landfill study area, Storrs, Connecticut. Values in parentheses next to the symbols indicate the elevation of open hole for that monitoring zone. Precipitation and packer-pressure data are shown below the hydrograph. Gaps in the record indicate missing or erroneous data that were omitted.

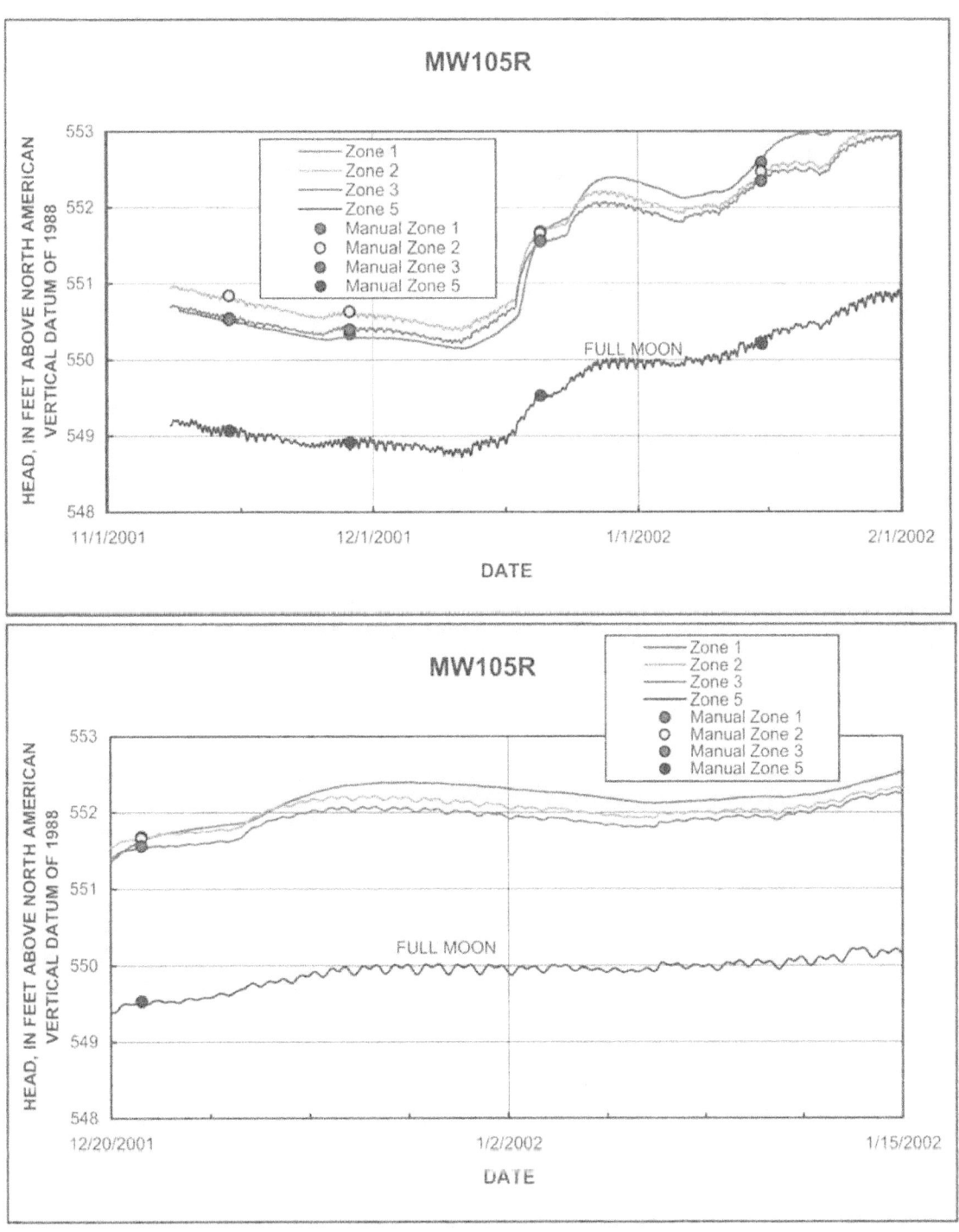

Figure 13. Continuous and manual water-level data for borehole MW105R in the UConn landfill study area, Storrs, Connecticut.

Table 8. Heads and transmissivity derived from heat-pulse flowmeter measurements in MW105R in the UConn landfill study area, Storrs, Connecticut, August 30, 1999.

[DZM, discrete-zone monitoring (system). Elevation is based on NAVD 88]

Depth, in feet below top of casing	Transmissive fracture in DZM zone	[1]Relative head, in feet	Water-level elevation, in feet	Transmissivity, in feet squared per day	Proportion of transmissivity, in percent
74.0	Zone 3	0.0	567.30	6.3	14
111.0-113.0	Zone 5	0.0	567.30	39.0	86

[1] Assumed same head because no ambient flow.

The expected effects of gravimetric forces for this location are superimposed on other natural fluctuations (fig.13). The variation in the amplitude of water-level change over a monthly cycle is related to the strength of the gravimetric forces computed over this time period for this location. The highest amplitude changes occur at the new and full moons, and the lowest amplitude responses occur between the new and full moon (quadrature). Figure 13 also demonstrates that the most transmissive zone in MW105R shows the highest amplitude of semi-diurnal change (0.12 ft), which is consistent with theoretical predictions (Hsieh and others, 1987). The Zone 2 and Zone 3 transducers show about half as much change in semi-diurnal amplitude (0.05 and 0.06 ft, respectively). The uppermost transducer shows little to no change in water-level response to tidal forces, which indicates that Zone 1 is unconfined and the water level in Zone 1 is unaffected by earth tides.

Borehole MW109R

Borehole MW109R is located on the hillside east of the landfill. The borehole was drilled using tricone roller bit drilling methods and was completed with 20 ft of 4.5-in-diameter casing and open below the casing to about 124 ft below land surface. Four 4-in-diameter packers were installed at center depths of 36.0, 50.0, 69.0, and 94.0 ft below the top of casing (fig. 8D). Five sampling and monitoring ports were installed at depths of 26.0, 44.0, 62.5, 75.0, and 110.0 ft below the top of casing. Below the casing, the DZM system isolates the five zones at elevations of 562.3-547.3; 545.3-533.3; 531.3-514.3; 512.3-489.3; and 487.3-455.3 ft above NAVD 88.

The hydrograph for MW109R is shown in figure 14. Over the period of record, the four lower zones (Zones 2, 3, 4, and 5) have similar heads. The head in Zone 1, however, is approximately 4 ft higher than the lower zones, indicating a strong downward driving potential.

Borehole MW109R shows strong seasonal fluctuations with water levels peaking in spring, declining through the summer and fall, and recharging in the winter and spring. Seasonal fluctuations associated with periods of minimal precipitation have as much as 15 and 17 ft of head decline, indicating the head response in this borehole is highly sensitive to precipitation (fig.14). Heads in all zones showed minor spikes in response to precipitation. Water levels in the overburden piezometer, MW109, were consistently higher than the water levels in the MW109R discrete zones, indicating a downward driving potential between the overburden and bedrock. During much of the period of record, however, MW109 was dry, and the water level was below an elevation of 566.47 ft.

Heat-pulse flowmeter measurements on July 20, 2000, in MW109R indicated that under ambient conditions, ground water entered the borehole at depths of 113 and about 75 ft and flowed upward and exited the borehole at about 62 ft below the top of casing. Concurrent with upflow in the lower zones of the borehole, water entered the borehole at depths of 26 and 44 ft below the top of casing and flowed downward, exiting the borehole at 62 ft below the top of casing. Heat-pulse flowmeter measurements indicate the fracture at 62 ft was the most transmissive fracture in the borehole. The heads in all other zones are consistent with the flow regime measured with the heat-pulse flowmeter. A comparison of open-hole head with DZM heads immediately after the installation of the DZM system on May 4,

2000, indicates Zone 3 had the lowest head, which is consistent with the flowmeter logs that identified the fracture in Zone 3 at 62 ft as the receiving fracture. The open-hole head measured on May 4, 2000, was nearly coincident with the heads in Zones 3, 4, and 5 (appendix 3E). Although Zone 3, which contains the fracture at 62 ft, was estimated to account for 80 percent of the transmissivity, the open-hole head on May 4, 2001, was still influenced by the higher heads in the other zones.

The heads and transmissivity computed from the two heat-pulse flowmeter profiles collected in July 1999 are shown in table 9. The model produced an SS error of 0.0301 (gal/min)^2. The head values appear to be reasonable for the summer season compared to the measured values in the DZM system collected in 2001 and 2002.

The transmissivity values determined with the flow-profiling model are different from the results of discrete-interval packer testing under pumping conditions. Table 10 compares the results of packer testing with the profiling. The packer test results were determined with a Cooper-Jacob straight-line solution (Haley and Aldrich, Inc. and others, 2002). The early-time data (less than about 10 min) yielded higher transmissivity values than the later time data, which fit to a lower transmissivity value. Because the flowmeter is used after the borehole has reached a quasi steady-state water level, which took about 15 min after the start of the test, the late time packer-test results probably are more comparable to the flowmeter results. The transmissivity estimates Zone 4 at a depth of 75 ft are much higher in the discrete interval tests than with the flow profiling.

Transducers were installed in Zones 1, 2, 3, and 4. The continuous water-level record showed fairly good correlation with the measured values. The continuous record identified more fluctuation in water levels than were recorded in the manual measurements. Only minor shifts in the transducer set depths were needed after the transducers were temporarily removed from the borehole for ground-water sampling.

All zones showed head increases that appear to be associated with precipitation. The heads in Zones 1 and 2 also showed larger responses to precipitation than Zones 3 and 4. Zones 2, 3, and 4 responded to the gravimetric forces and a well-defined semi-diurnal pattern was observed in each. During full- and new-moon conditions, the amplitude of the daily water-level changes was about 0.05 ft in Zones 2, 3, and 5. The uppermost zone, Zone 1, does not show gravimetric tidal fluctuation, indicating it is unconfined.

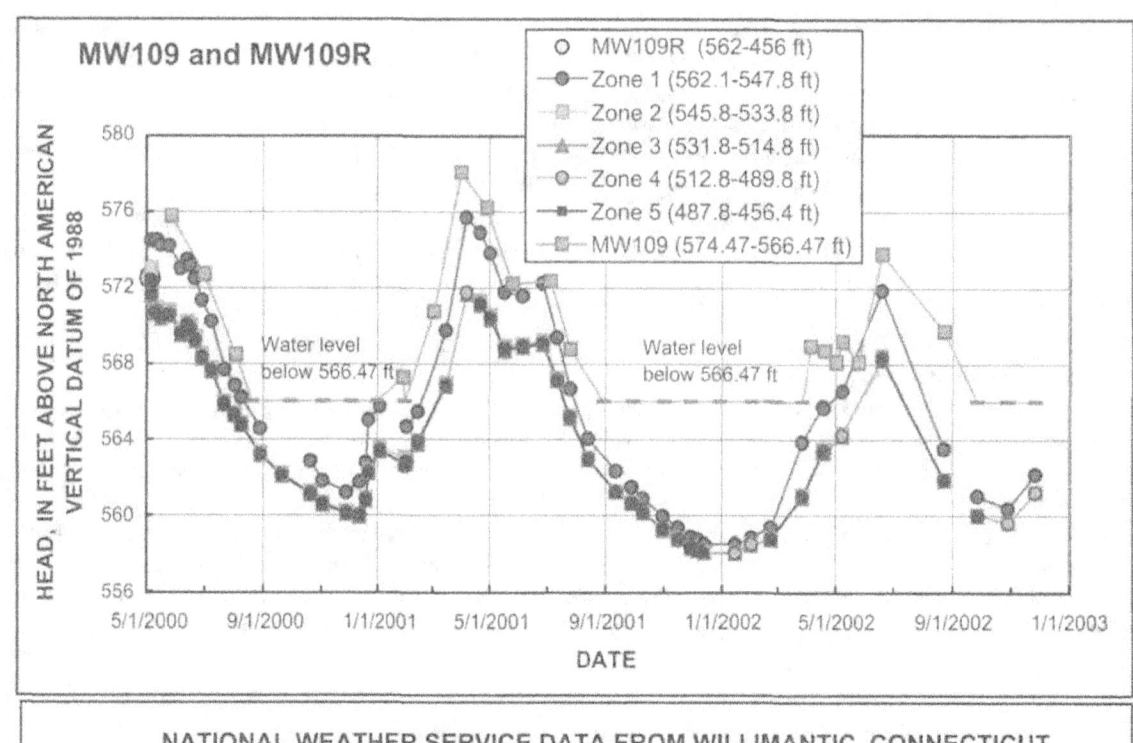

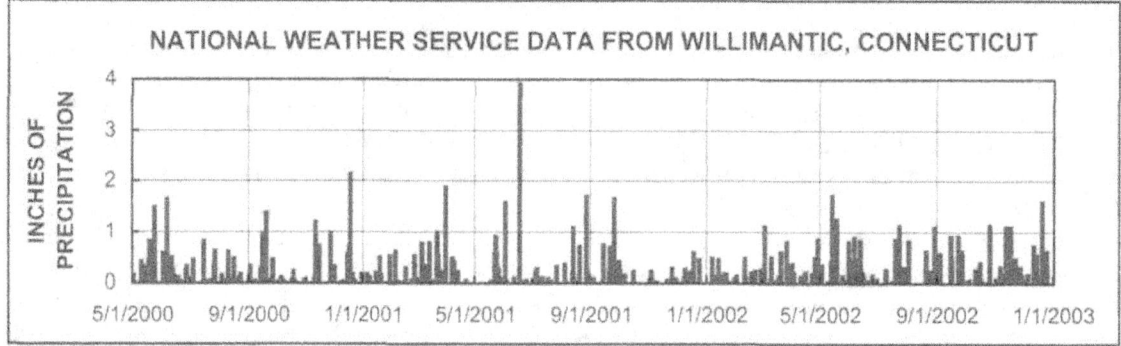

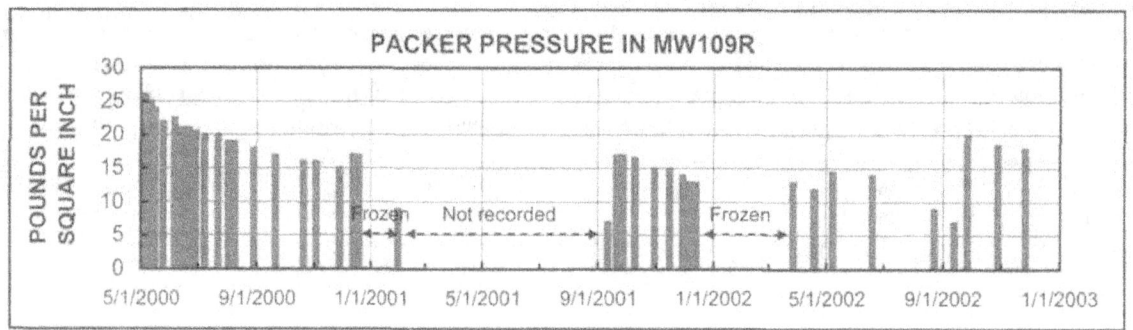

Figure 14. Hydrographs for MW109R and MW109 in the UConn landfill study area, Storrs, Connecticut. Values in parentheses next to the symbols indicate the elevation of open hole for that monitoring zone. Precipitation and packer-pressure data are shown below the hydrograph. Gaps in the record indicate missing or erroneous data that were omitted.

Table 9. Heads and transmissivity derived from heat-pulse flowmeter measurements in MW109R in the UConn landfill study area, Storrs, Connecticut, July 20, 1999.

[DZM, discrete-zone monitoring (system). Elevation is based on NAVD 88]

Depth, in feet below top of casing	Transmissive fracture in DZM zone	[1]Relative head, in feet	Water-level elevation, in feet	Transmissivity, in feet squared per day	Proportion of transmissivity, in percent
25.5	Zone 1	0.7	564.59	17.7	4
44.0	Zone 2	-0.8	563.09	19.2	12
62.0	Zone 3	-1.6	562.29	198.0	79
75.0	Zone 4	0.6	564.49	17.8	3
113.0	Zone 5	0.0	563.89	18.4	1

[1] Relative head is the magnitude of head for the zone relative to the head of the lowermost zone.

Table 10. Transmissivity values estimated from flow-profiling modeling and discrete-interval packer testing in MW109R in the UConn landfill study area, Storrs, Connecticut, July 20, 1999.

[Haley and Aldrich, Inc. and others, 2002]

Depth, in feet below top of casing	Flow profiling and modeling transmissivity, in feet squared per day	[1]Late time discrete-interval packer testing transmissivity, in feet squared per day	[2]Early time discrete-interval packer testing transmissivity, in feet squared per day
62	198.0	47	210
75	17.9	52	400

[1]Transmissivity determined with Cooper-Jacob straight-line method, where the line was fit to late-time water-level data collected about 10 minutes after initiating pumping.

[2]Transmissivity determined with Cooper-Jacob straight-line method, where the line was fit to early-time water-level data collected less than 10 minutes after initiating pumping.

Borehole MW121R

Borehole MW121R is located approximately 550 ft west of the landfill, 320 ft northwest of the former chemical-waste disposal pits, and about 150 ft north-northeast of MW202R (fig. 1). The borehole was drilled using a tricone roller bit and was completed with 11 ft of 5-in-diameter steel casing and open below the casing to about 128 ft below land surface. One 4-in-diameter packer was installed on April 20, 2000, at a center depth of 56.5 ft below the top of casing (fig. 8E). Two monitoring and sampling ports were installed at depths of 53.0 and 65.5 ft below the top of casing. Below the casing, the DZM system isolates two zones at elevations of 577.8-533.3 and 531.3-461.8 ft above NAVD 88. The deeper zone appears to be the more transmissive of the two zones, based on the comparison of the open-hole water level collected on April 18, 2000, to discrete-zone water levels collected on April 20, 2000.

Over the period of record, the head in the upper zone is consistently higher than the heads in the lower zone, indicating a downward driving potential (fig. 15). Borehole MW121R shows strong seasonal fluctuations with water levels peaking in spring, declining through the summer and fall, and recharging in the winter and spring. From spring to winter 2000, the water levels in the upper zone showed as much as a 14.5-ft decline in response to drought, and there was a 10-ft decline in the lower zone. During the summer and fall dry seasons, the driving potential was still downward, although it was less than in the spring. The water levels in both zones respond to precipitation; however, the response in the lower zone is more subdued than in the upper zone. During a period of recharge, mid-December 2000 through June 2001, the driving potential was downward, which indicates seasonal recharge. During summer and fall dry periods, the heads in both zones converge toward one another, indicating only a small downward driving potential that is much less than the driving potential during periods of recharge.

The heads in the upper zone of MW121R are slightly higher than the heads in MW122R. The general seasonal pattern in the hydrograph for MW121R is similar to MW122R and MW109R. The magnitude of the heads during the spring is as high as the peak heads in MW109R; however, the magnitude of the seasonal low is not as low as in MW109R. Borehole MW121R is interpreted as being in a recharge zone.

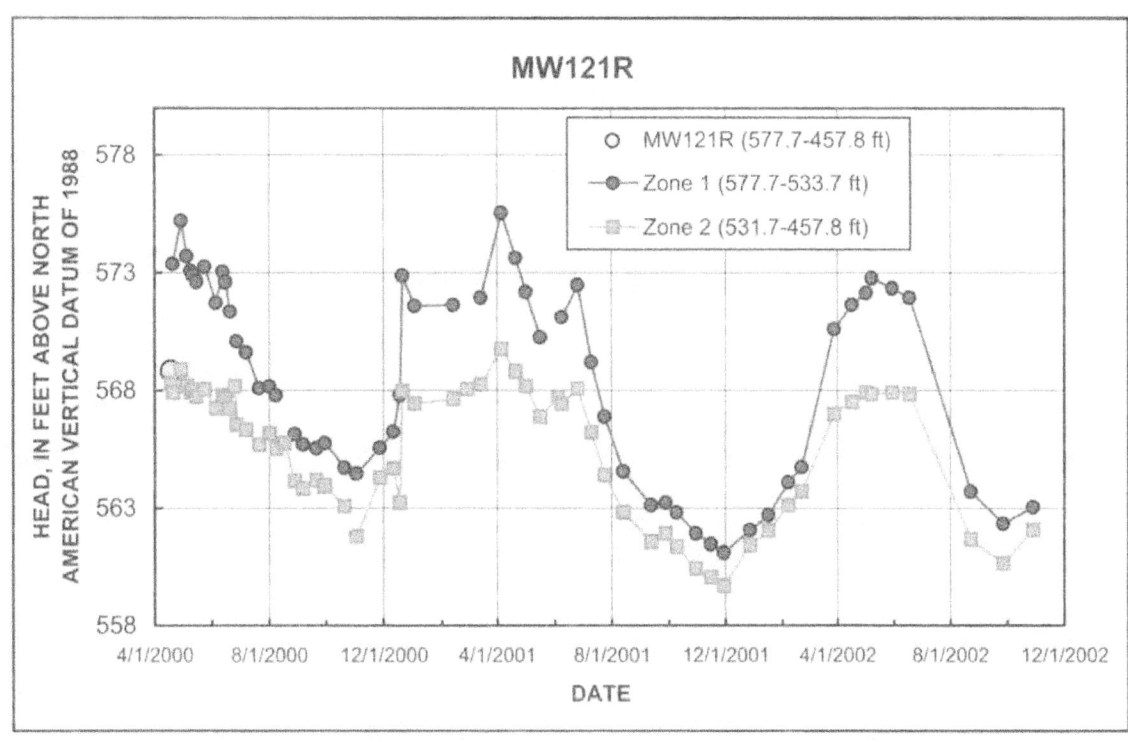

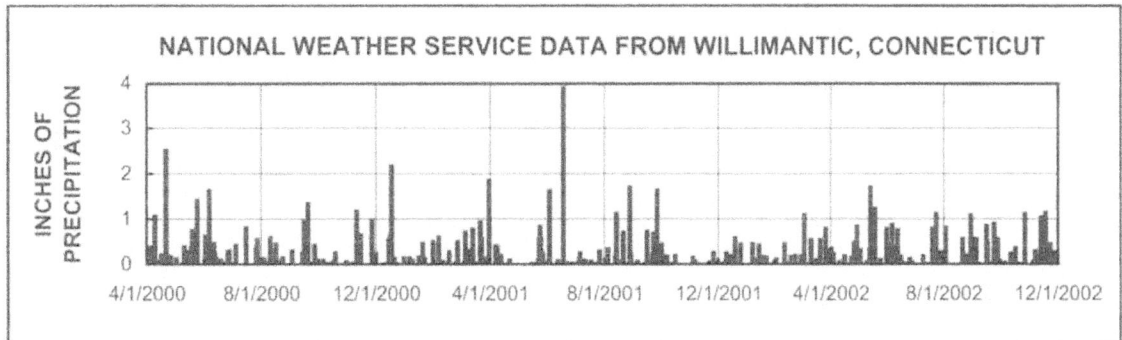

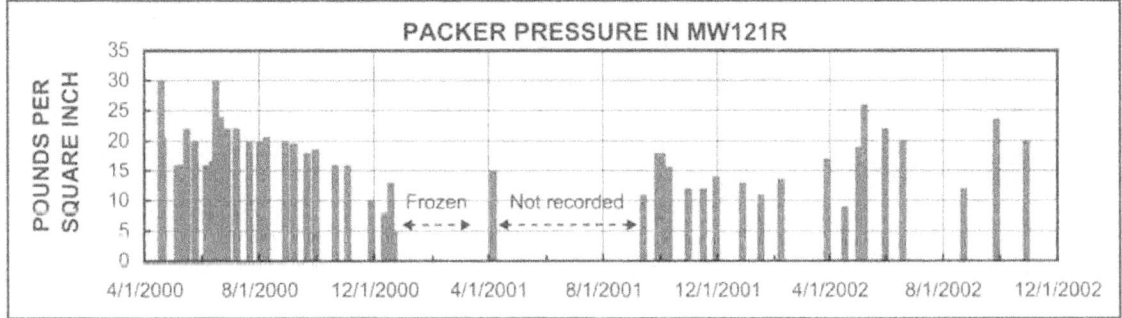

Figure 15. Hydrograph for MW121R in the UConn landfill study area, Storrs, Connecticut. Values in parentheses next to the symbols indicate the elevation of open hole for that monitoring zone. Precipitation and packer-pressure data are shown below the hydrograph. Gaps in the record indicate missing or erroneous data that were omitted.

Borehole MW122R

Borehole MW122R is located 500 ft west of the landfill and 50 ft west of the former chemical-waste disposal pits (fig. 1). The borehole was drilled using air-hammer rotary methods and was completed with 11.5 ft of 6-in-diameter casing and open below the casing to about 125 ft below land surface. Two packers were installed on June 27, 2000, at center depths of 59.5 and 71.0 ft below the top of casing (fig. 8F). Sampling ports were installed adjacent to fractures that were identified as transmissive (Johnson and others, 2002) at depths of 57.0, 61.5, and 80.0 ft below the top of casing. Below the steel casing, the DZM system isolates three zones at elevations 569.9-522.9; 520.9-511.4; and 509.4-454.4 ft above NAVD 88.

The hydrograph for MW122R (fig. 16) indicates that the open-hole head on June 27, 2000, coincides with the head in Zone 2, indicating this is the most transmissive zone. Water levels in Zones 1 and 3 coincide with one another over the entire period of record. Pumping in Zone 1 produced an immediate drawdown in Zone 1, but not in Zone 3, indicating that these zones are not connected directly through cross connection in the borehole, which would indicate packer failure. During periods of recharge from mid-December 2000 to mid-August 2001, Zone 2 consistently showed lower heads than Zones 1 and 3. During the summer and fall dry seasons in 2001 and 2002, the heads in all three zones were the same, indicating vertical equipotentials and no vertical flow in the area. A similar pattern was observed in nearby boreholes MW121R, MW203R, and MW204R.

There were insufficient data to compute the heads from heat-pulse flowmeter profiling in MW122R. There was no measurable vertical flow under ambient conditions, and the water levels did not equilibrate under pumping conditions

Transducers were installed in Zones 1 and 2. The continuous record showed fairly good correlation with the measured values. The continuous record identified more fluctuation in water levels than were recorded in the manual measurements. Only minor shifts in the transducer set depths were needed after the transducers were temporarily removed from the borehole for ground-water sampling.

Zones 1 and 2 show abrupt increases in water level that appear to be caused by precipitation events (appendix 3F). The continuous record showed more variation in the water levels in Zone 1 than in the deeper Zone 2, which showed a strong response to the gravimetric forces and a well-defined semi-diurnal pattern. Under full- and new-moon conditions, the amplitude of the daily water-level changes was 0.05 ft in Zone 2. The uppermost zone, Zone 1, does not show gravimetric tidal fluctuation, indicating it is unconfined.

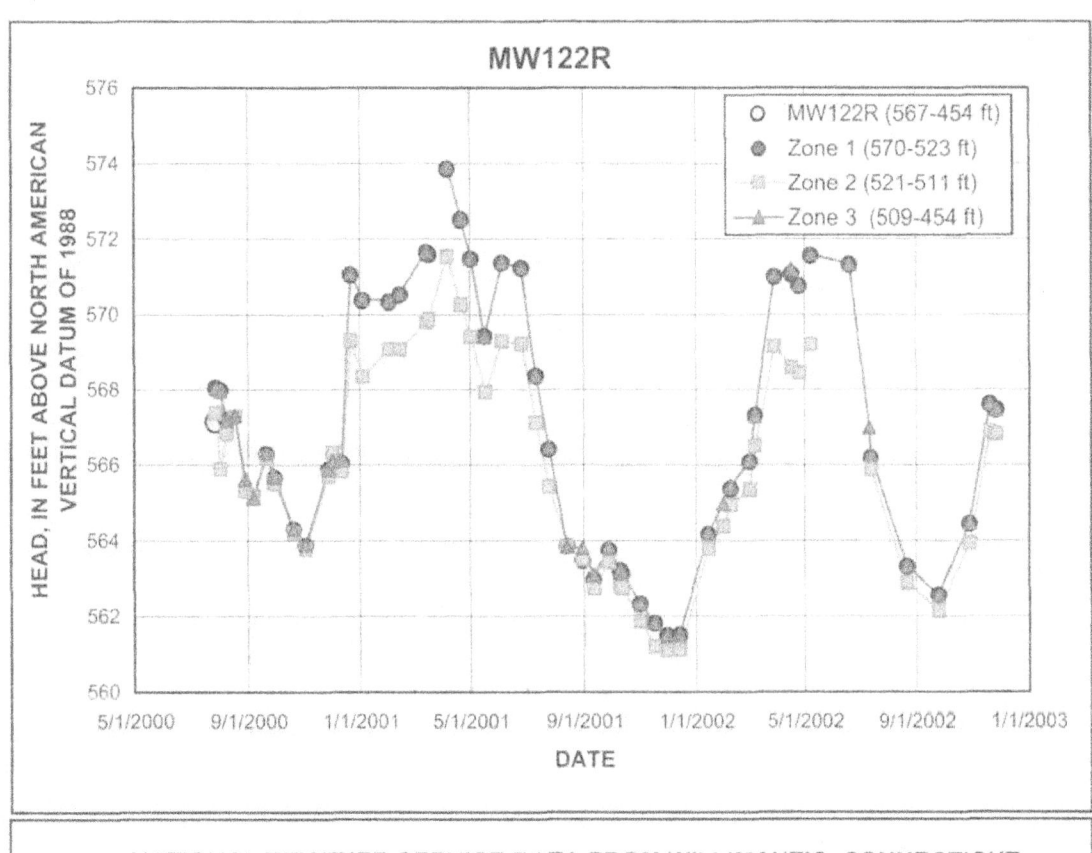

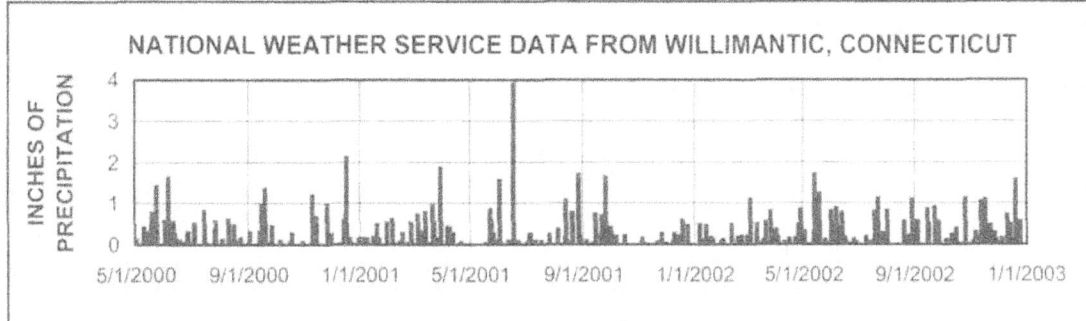

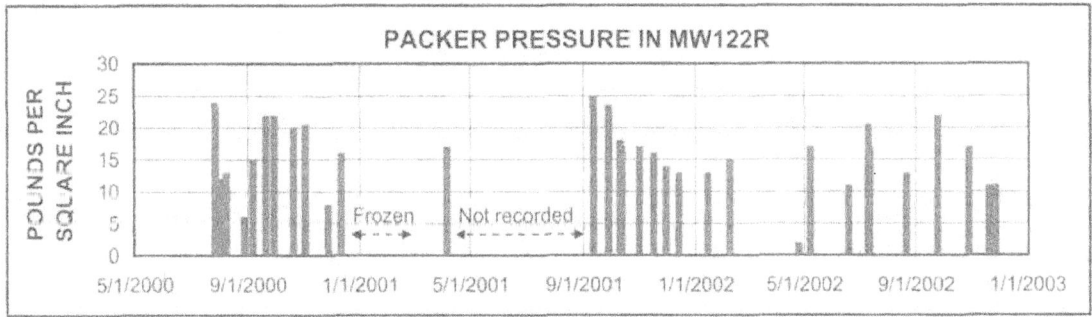

Figure 16. Hydrograph for MW122R in the UConn landfill study area, Storrs, Connecticut. Values in parentheses next to the symbols indicate the elevation of open hole for that monitoring zone. Precipitation and packer-pressure data are shown below the hydrograph. Gaps in the record indicate missing or erroneous data that were omitted.

Borehole MW201R

Borehole MW201R is located approximately 350 ft south of MW105R and about 500 ft southwest of the landfill (fig. 1). The borehole was drilled using air-hammer rotary methods and was completed with 15.2 ft of 6-in-diameter casing and open below the casing to about 124 ft below land surface. A total of three packers was installed on April 5, 2001, at center depths of 50.0, 70.0, and 90.0 ft below the top of casing (fig. 8G). Sampling and monitoring ports were installed at depths of 38.0, 60.0, 78.0, and 97.0 ft below the top of casing. Below the casing, the DZM system isolates four zones at elevations of 539.0-505.2; 503.2-485.2; 483.2-465.2; and 463.2-428.6 ft above NAVD 88. Transmissive fractures were identified at depths of about 38 and 60 ft below the top of casing. The fractures identified at depths of about 76 and 97 ft, were not identified as hydraulically active with the heat-pulse flowmeter. In addition, no ambient flow was measured in the borehole with the heat-pulse flowmeter in August and September 2000.

The hydrograph for MW201R is shown in figure 17. Over the entire period of record, the driving potential between the two most transmissive zones (Zones 1 and 2) was consistently upwards. During periods of recharge, MW201R showed strong upward driving potentials. On June 6, 2000, when MW201R was drilled, the ambient open-hole head was approximately 0.5 ft above the land surface. In the spring of 2000 and 2001, after the DZM system was installed, water flowed from all four zones. On April 24, 2001, the top of the CMT was redesigned to prevent potential cross contamination from overflow at the top of the borehole. Small v-shaped notches were cut into the side of each channel so that the overflowing water would discharge to a collection system that diverted water away from other channels and the 6-in-diameter steel casing. The overflowing water discharged to the land surface. The hydraulic heads in Zones 2 and 3 coincide over most of the record, and those heads are consistently higher than the head in Zone 4.

Differential head tests made with the BAT[3] system during September 2000 had fairly good agreement with some minor differences (table 11). The BAT[3] indicated the head for the fracture at a depth of 38 ft was 550.97 ft above NAVD 88 and the head for the fracture at 60 ft was 551.64 ft above NAVD 88. These results correlate with the results of the DZM system. If only the head differentials during the tests are considered, however, then the results might appear inconclusive. The differential head between the test zone at 38 ft and the lower zone had an upward driving potential. The differential head between the upper zone and the test zone at a depth of 60 ft was downward. This discrepancy is either because of measurement error, or because the heads in the zones above and below the test interval are transmissivity-weighted heads influenced by all the heads intersecting the borehole above or below the test interval.

Because there was no ambient flow, there were insufficient flowmeter data to model the head and the transmissivity using flow profiles.

MW201R is interpreted to be located in a discharge zone along the southwest drainage from the landfill. This interpretation is consistent with the observation that the heads in Zones 2 and 3 are consistently higher than the head in Zone 1, and the head in Zone 1 is higher than the head observed in piezometer 13, which is completed about 10 ft into the bedrock.

Table 11. Differential heads in MW201R from BAT[3] testing in the UConn landfill study area, Storrs, Connecticut.

[TOC, top of casing, in feet above NAVD 88; *, no data; --, no value could be computed; upper, middle, and lower zones isolated by packers]

Well	Date of test	[1]Depth of fracture, in feet	TOC	Hydraulic head, in feet above NAVD 88			[2]Head differential, in feet		
				Upper	Middle	Lower	Upper-Mid	Mid-Lower	Upper-Lower
MW201R	9/28/2000	38	554.22	551.34	550.97	551.59	0.37	-0.62	-0.25
	9/29/2000	60		551.87	551.64	552.12	0.23	-0.49	-0.25
	9/28/2000	76		551.59	*	546.23	--	--	5.36
	9/29/2000	97		551.57	551.55	549.24	0.03	2.30	2.33

[1]Fracture is located within the middle zone.

[2]Head differential – positive number denotes downward driving potential, negative denotes upward driving potential.

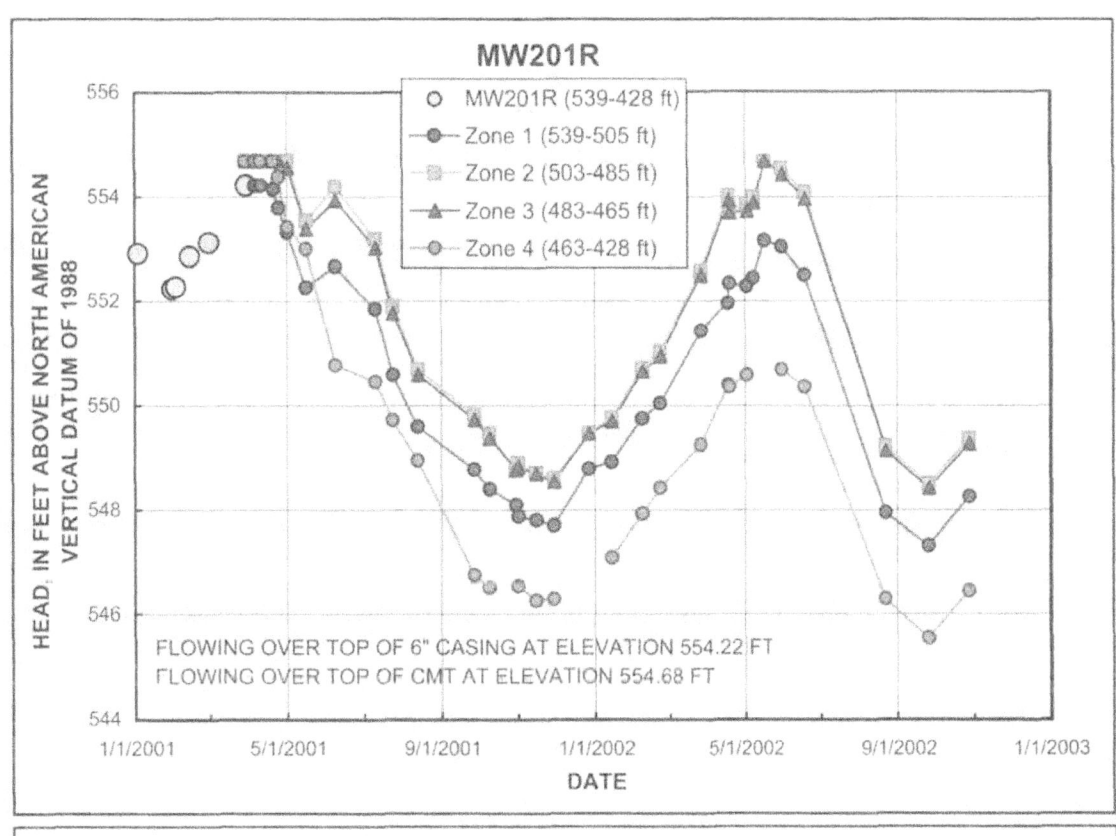

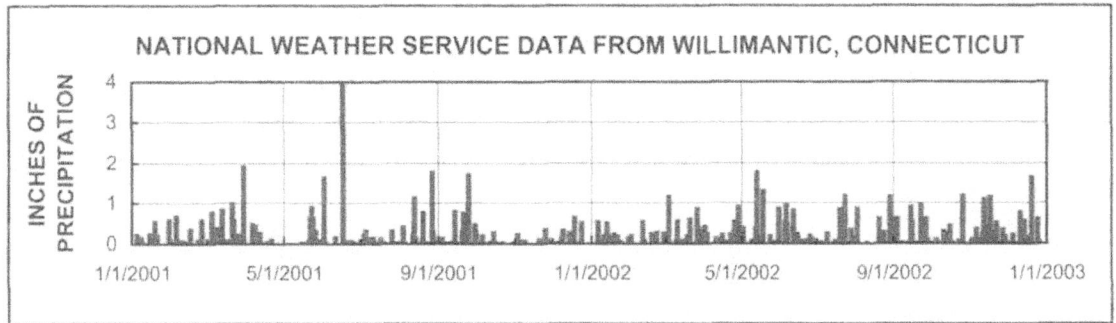

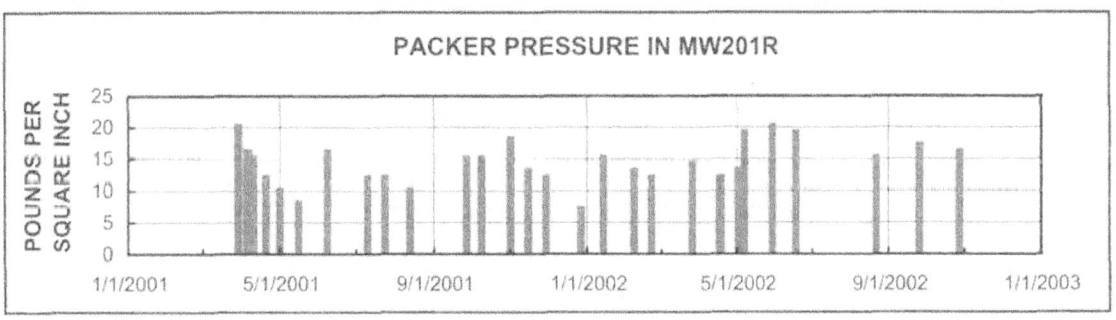

Figure 17. Hydrograph for MW201R in the UConn landfill study area, Storrs, Connecticut. Values in parentheses next to the symbols indicate the elevation of open hole for that monitoring zone. Precipitation and packer-pressure data are shown below the hydrograph. Gaps in the record indicate missing or erroneous data that were omitted.

Borehole MW202R

Borehole MW202R is located about 400 ft west of the former chemical-waste disposal pits and about 500 ft east of Hunting Lodge Road (fig. 1). MW202R was drilled using air-hammer rotary methods and was completed with 13.5 ft of 6-in-diameter casing. Below the casing, the borehole is completed as a 6-in-diameter hole to a depth of 125 ft below land surface. Two packers were installed on March 29, 2001, at center depths of 38.0 and 100.0 ft below the top of casing (fig. 8H). Ports were installed at depths of 30.0, 45.0, and 107.0 ft below the top of casing. Below the casing, the DZM system isolates three zones at elevations of 568.8-545.3; 543.3-483.3; and 481.3-455.4 ft above NAVD 88. Transmissive fractures were identified with the heat-pulse flowmeter at 30.9 ft and 107.0 ft below the top of casing. The open-hole head collected on March 29, 2001, is about half way between the heads in Zone 1 and Zone 3, indicating that they are both transmissive.

The seasonal fluctuations from the spring peak in water level to the lowest water levels in winter show a decline of about 10 ft in all of the zones. The hydrograph for MW202R indicates the water-level response to precipitation is fairly gradual, except for a large rise in water levels after precipitation on June 17, 2001 (fig. 18). There were downward driving potentials from Zone 1 to Zone 2 and from Zone 2 to Zone 3 over most of the period of record, except from January 1 to March 1, 2002. During this period in the winter of 2002, Zone 1 and Zone 2 converged, even though the packers reportedly maintained inflation. This may represent a failed gage and a drop in packer pressure, or it may be accurate. It was not a complete failure of the packers, or water levels in all zones would have converged to the same water level.

A comparison of the elevation of the heads determined with the BAT[3] system to the heads collected in the DZM system in October 2001 indicates fairly good agreement (table 12). The direction of potential flow was the same for the BAT[3] method and the DZM system, but the differential heads between the BAT[3] data for the two most transmissive zones were greater than the differential heads measured between Zone 1 and Zone 3 in the DZM system for the same time of year.

In comparison with nearby boreholes, the heads in the upper zone of the DZM system in MW202R were higher than the heads in all zones in MW122R, MW204R, and MW203R. Also, the heads in all zones of MW202R are consistently higher than the open-hole head in MW302R. These results are consistent with the interpretation that MW202R is in a recharge zone.

Table 12. Differential heads in MW202R from BAT[3] testing in the UConn landfill study area, Storrs, Connecticut.

[TOC, top of casing, in feet above NAVD 88; upper, middle, and lower zones isolated by packers]

Well	Date of test	[1]Depth of fracture, in feet	TOC	Hydraulic head, in feet above NAVD 88			[2]Head differential, in feet		
				Upper	Middle	Lower	Upper-Mid	Mid-Lower	Upper-Lower
MW202R	10/4/2000	31	582.33	566.43	567.21	564.56	-0.78	2.65	1.87
	10/5/2000	107		567.21	558.57	564.17	8.64	-5.60	3.04

[1]Fracture is located within the middle zone.

[2]Head differential – positive number denotes downward driving potential, negative denotes upward driving potential.

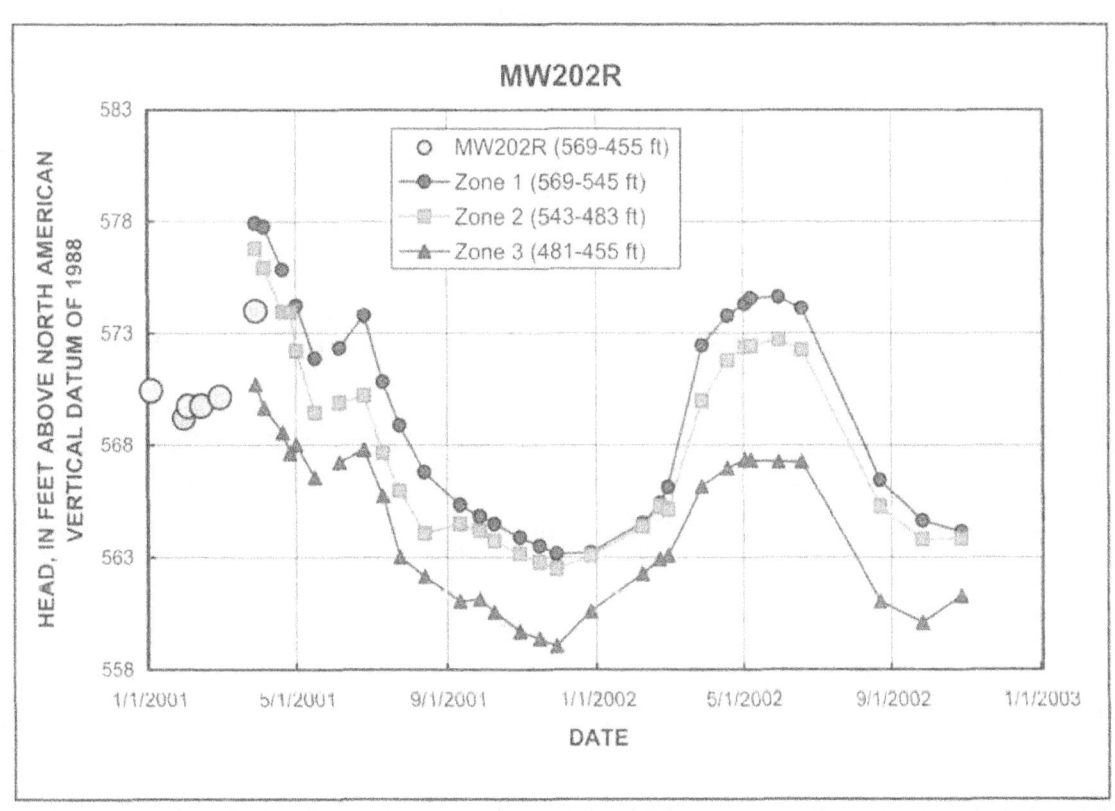

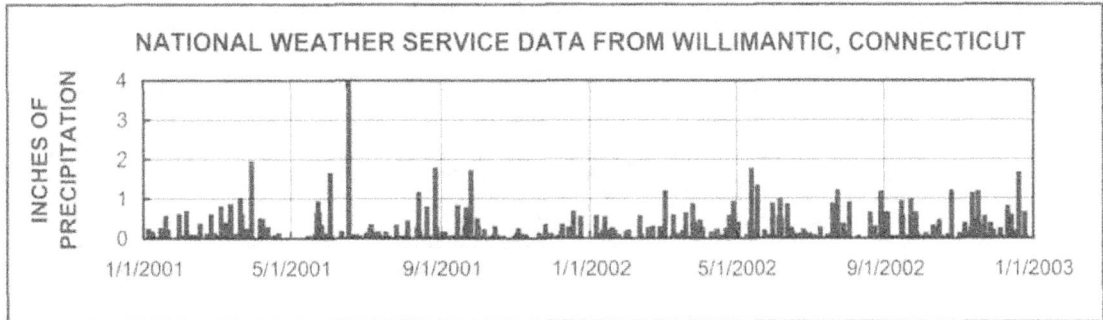

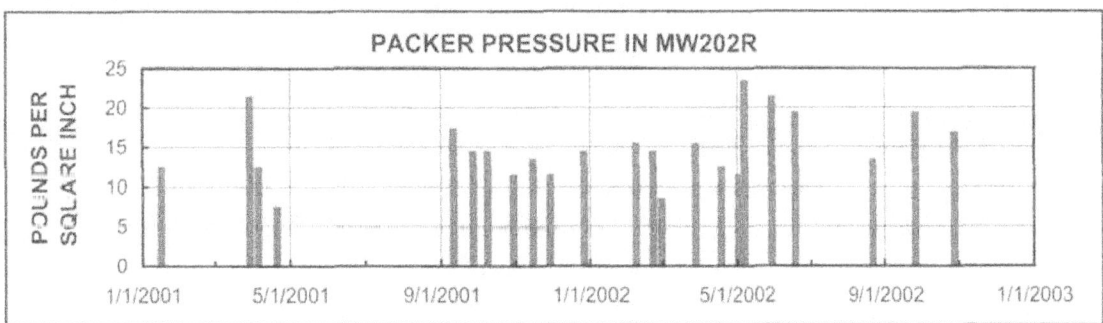

Figure 18. Hydrograph for MW202R in the UConn landfill study area, Storrs, Connecticut. Values in parentheses next to the symbols indicate the elevation of open hole for that monitoring zone. Precipitation and packer-pressure data are shown below the hydrograph. Gaps in the record indicate missing or erroneous data that were omitted.

Borehole MW203R

Borehole MW203R is located about 200 ft west of the landfill within 100 ft of the former chemical-waste disposal pits (fig. 1). MW203R was drilled using air-hammer rotary methods and was completed with 14.4 ft of 6-in-diameter casing and open below the casing to about 124 ft below land surface. Two packers were installed on March 20, 2001, at center depths of 44.0 and 89.0 ft below the top of casing (fig. 8I). Fractures were identified by borehole-geophysical methods at depths of 32 ft and above; at 62, 65, and 68 ft; and at 95 and 108 ft (Johnson and others, 2005). Monitoring and sampling ports were installed at depths of 32, 65, and 103 ft below the top of casing. Below the casing, the DZM system isolates three zones at elevations of 562.5-533.9; 531.9-488.9; and 486.9-451.9 ft above NAVD 88. Only the fractures in the upper zone were identified as transmissive with the heat-pulse flowmeter and the BAT[3] equipment. The fracture at a depth of 65 ft below the top of casing could not sustain a pumping rate of 0.15 gal/min with the BAT[3] equipment. The open-hole water level on March 20, 2001, was almost mid-way between Zone 1 and Zone 2, indicating that Zone 2 also has some transmissivity and some influence on the open-hole water level.

The hydrograph for MW203R is shown in figure 19. During periods of seasonal recharge there are downward driving potentials from Zone 1 to Zone 2 and from Zone 2 to Zone 3. During the summer, from June 5 to September 27, 2001, the heads in all three zones converged, while the packers reportedly maintained inflation. This seasonal pattern is similar to MW121R and MW122R; however, the water levels from July 1 to August 1, 2001, probably are erroneous. Three conditions suggest there was a packer failure. First, the water levels converge to the highest head in the borehole in Zone 1 rather than declining to the lower heads, as would be expected if this were caused solely from a lack of precipitation. Second, the packer inflation measurements read off of the gage are anomalously high; measurements were reportedly 45 psi above ambient. Third, the water levels diverged just after the gage was replaced and the packers were pressurized.

A comparison of the head measurements between the DZM system and the BAT[3] shows a fairly poor correlation. Both methods identified downward driving potentials between the zones in the borehole. The BAT[3] system measured only small head differentials between the upper and middle zones for the tests conducted at depths of 32 ft and 65 ft (table 13). The hydrograph for the DZM system in October 2001 indicates a strong downward driving potential of about 6 ft between Zone 1 and Zone 2, and a differential head of 1.6 ft between Zone 2 and Zone 3.

Borehole MW203R is located just south of the north-south ground-water divide in the area of the former chemical-waste disposal pits. The downward driving potentials identified in MW203R during periods of recharge are consistent with the placement of this borehole in a local recharge area. The heads in MW203R are consistently lower than the heads in MW104R, MW122R, and MW204R, and higher than the heads in MW105R and MW201R. These results indicate that MW203R is at an intermediate location along a southern flowpath extending from the ground-water divide south towards the intermittent drainage southwest of the landfill.

Table 13. Differential heads in MW203R from BAT[3] testing in the UConn landfill study area, Storrs, Connecticut.

[TOC, top of casing, in feet above NAVD 88; upper, middle, and lower zones isolated by packers]

Well	Date of test	[1]Depth of fracture, in feet	TOC	Hydraulic head, in feet above NAVD 88			[2]Head differential, in feet		
				Upper	Middle	Lower	Upper-Mid	Mid-Lower	Upper-Lower
MW203R	10/13/2000	32	576.9	564.71	561.04	561.06	3.67	[3]-0.02	3.65
		65		562.16	560.50	554.50	1.66	6.00	7.66

[1]Fracture is located within the middle zone.

[2]Head differential – positive number denotes downward driving potential, negative denotes upward driving potential.

[3]The differential head is less than the error range for the measurements.

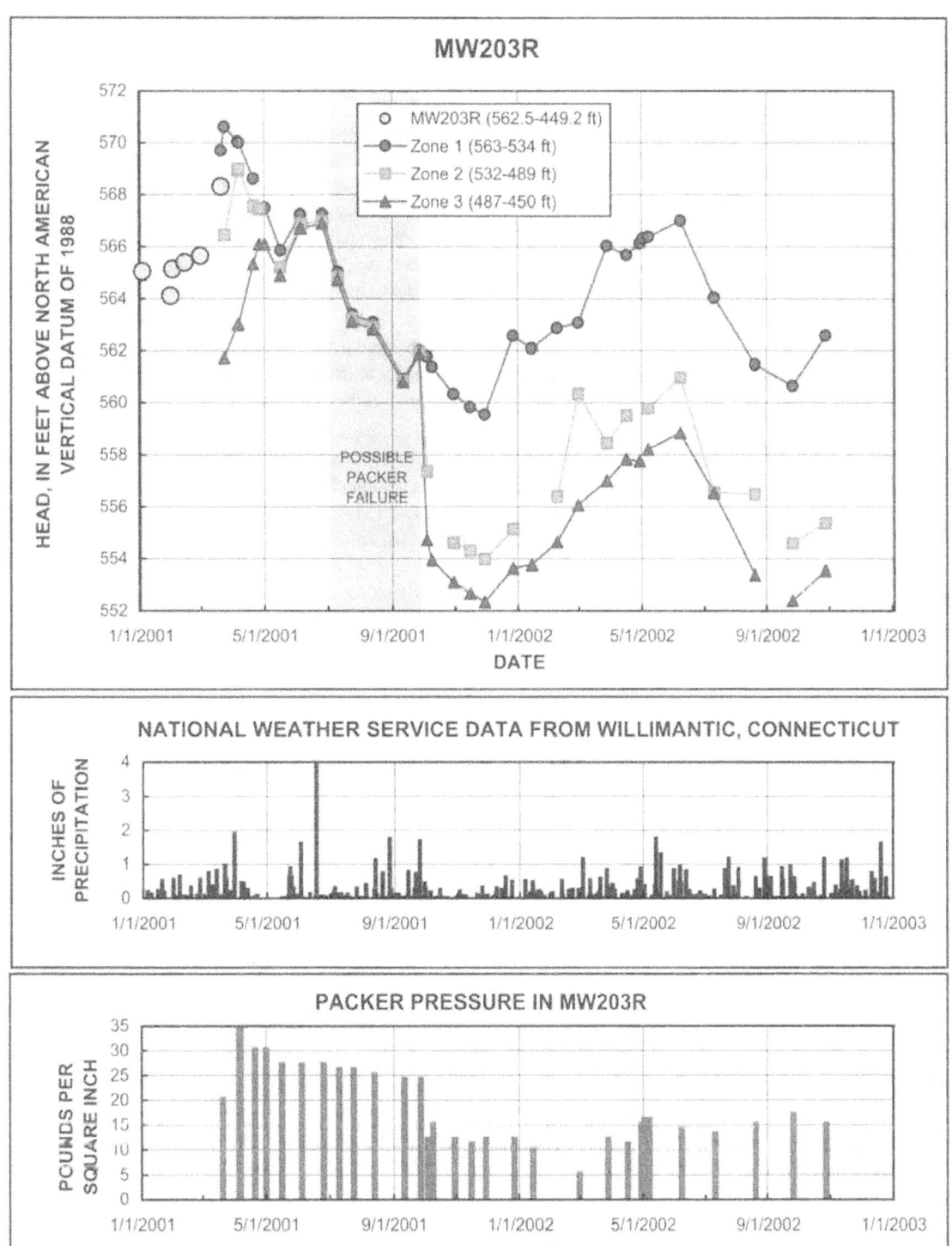

Figure 19. Hydrograph for MW203R in the UConn landfill study area, Storrs, Connecticut. Values in parentheses next to the symbols indicate the elevation of open hole for that monitoring zone. Precipitation and packer-pressure data are shown below the hydrograph. Gaps in the record indicate missing or erroneous data that were omitted.

Borehole MW204R

Borehole MW204R is located about 100 ft west of the landfill and about 200 ft north of the former chemical-waste disposal pits (fig. 1). MW204R was drilled using air-hammer rotary methods and was completed with 13.1 ft of 6-in-diameter casing and open below the casing to 126 ft below land surface. Two packers were installed on March 23, 2001, at center depths of 31.0 and 69.0 ft below the top of casing (fig. 8J). Below the casing, the DZM system isolates three zones at elevations of 562.2-545.3; 543.3-507.3; and 505.3-447.6 ft above NAVD 88. Fractures were identified by borehole geophysical methods at 21 and 22 ft; at 40, 45, 55, and 58 ft; and at 75, 87, and 90 ft (Johnson and others, 2005). Monitoring and sampling ports were installed at depths of 19.0, 50.0, and 80.0 ft below the top of casing. Only the fractures in the upper zone (from 13 to 25 ft) were identified as transmissive with the heat-pulse flowmeter and the BAT^3 equipment. The fracture at 45 ft below the top of casing could not sustain a pumping rate of 0.15 gal/min with the BAT^3 equipment. The open-hole water level on March 23, 2001, was at an elevation of 573.21 ft. The water levels collected after the DZM system was installed on March 23, 2001, had not equilibrated. The water levels on April 4, 2001, indicate that Zone 1 has the highest head and is most similar to the open-hole head.

The hydrograph for MW204R is shown in figure 20. Over the entire period of record, the heads in Zone 2 and Zone 3 are the same. During periods of seasonal recharge, there are downward driving potentials from Zone 1 to Zone 2. During the summer, from June 25 to September 27, 2001, the heads in all three zones converged, while the packers reportedly maintained inflation. From November 2001 to the end of the period of record, there were downward potentials between Zones 1 and 2.

A comparison of heads measured in the DZM system and heads measured with the BAT^3 system has mixed results. The discrete-interval water levels collected with the DZM system in October 2000 compare well with the BAT^3 results collected in the most transmissive part of the borehole, the uppermost zone. Both methods show an overall downward driving potential. The absolute isolated heads measured in BAT^3 indicate that the water levels in the fractures from depths of 13 to 29 ft and at 45 ft are essentially the same, within the measurement resolution of the transducers (table 14). In addition, the BAT^3 indicates that the heads in fractures in Zones 1 and 2 are similar and are higher than the head in Zone 3. The long-term record for the DZM system indicates, however, that the heads in Zone 2 and Zone 3 are similar, and they are always lower than the head in Zone 1.

The heads in the uppermost zone of MW204R are occasionally higher than the heads in the uppermost zone in MW104R, indicating that there may be driving potentials southward from MW204R towards the former chemical-waste disposal pits. Also, the heads in MW204R are consistently higher than the heads in MW103R and MW101R to the north. Borehole MW204R is located just north of the north-south groundwater divide, which is in the area of the former chemical-waste disposal pits. The downward driving potentials identified in MW204R during periods of recharge and the driving potentials with nearby boreholes are consistent with the conceptual placement of this borehole in a local recharge area.

Table 14. Differential heads in MW204R from BAT^3 testing in the UConn landfill study area, Storrs, Connecticut.

[TOC, top of casing, in feet above NAVD 88; upper, middle, and lower zones isolated by packers]

Well	Date of test	[1]Depth of fracture, in feet	TOC	Hydraulic head, in feet above NAVD 88			[2]Head differential, in feet		
				Upper	Middle	Lower	Upper-Mid	Mid-Lower	Upper-Lower
MW204R	10/11/2000	[3]21, 22	575.34	[3]565.63	565.63	563.98	[2]0.0	1.65	1.65
	10/11/2000	45		565.75	566.03	563.87	-0.28	2.16	1.88
	10/10/2000	76		565.62	563.91	556.10	1.71	7.81	9.52

[1]Fracture is located within the middle zone.

[2]Head differential – positive number denotes downward driving potential, negative denotes upward driving potential.

[3]Upper and middle zone were the same in this test, as the upper packer was not inflated. The test zone included multiple fractures from 13 to 29 ft. The most prominent were at 21 and 22 ft below the top of casing.

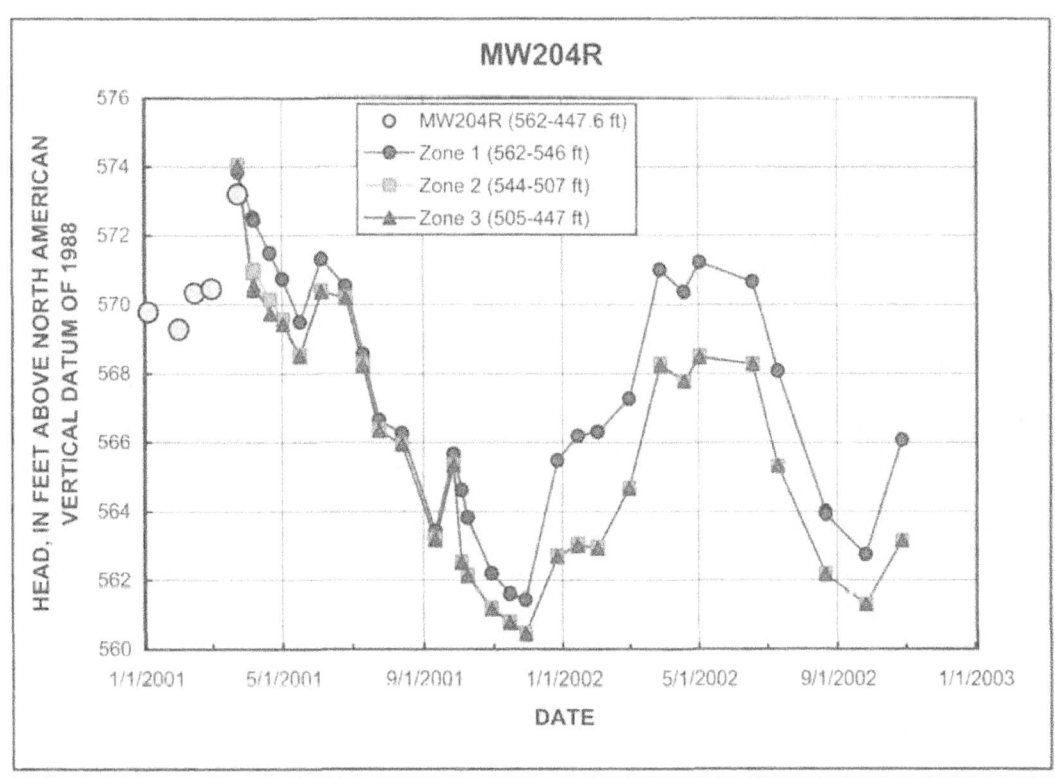

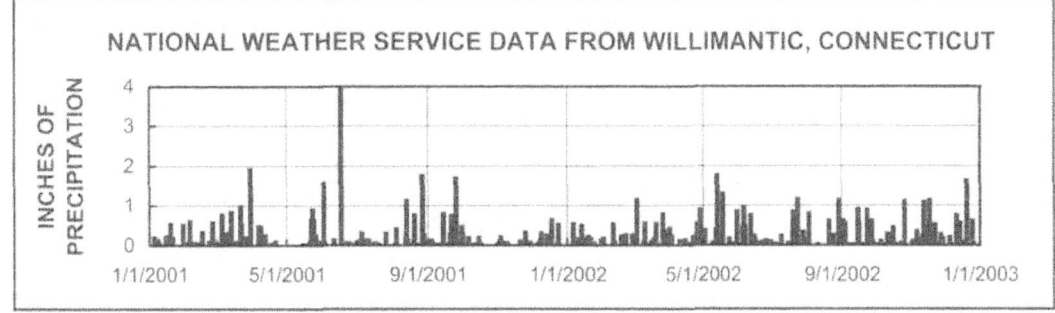

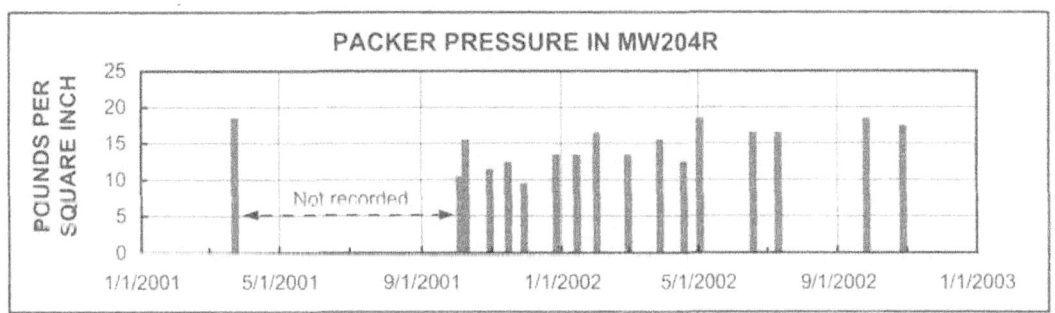

Figure 20. Hydrograph for MW204R in the UConn landfill study area, Storrs, Connecticut. Values in parentheses next to the symbols indicate the elevation of open hole for that monitoring zone. Precipitation and packer-pressure data are shown below the hydrograph. Gaps in the record indicate missing or erroneous data that were omitted.

Borehole MW302R

Borehole MW302R is located west about 1000 ft west of the landfill and about 200 ft west of MW202R (fig. 1). The borehole also is about 2,400 ft east of a domestic supply well that services multiple dwellings along Carriage House Road. MW302R was drilled to a depth of about 301 ft below land surface using dual-rotary methods that advanced a 10-in-diameter spin casing through the overburden simultaneously with the advancement of an air-rotary hammer. An 8-in-diameter poly-vinyl chloride (PVC) casing was installed to a depth of 127 ft, and a 6-in-diameter borehole was drilled to a depth of 303 ft below the top of the PVC casing using air-hammer rotary methods. Because of the depth of the borehole, a CMT system was not installed in MW302R.

The hydrograph constructed from manual open-hole water-level measurements is shown in figure 21. Although somewhat subdued, the fluctuation pattern in this borehole is the same as that of the other boreholes with DZM completions. The hydrograph shows seasonal fluctuations with a maximum head difference of about 6 ft (fig. 21). The cycle for 2001 and 2002 shows a decline in late fall and a general rise in water level through spring and into early summer.

Although no DZM system was installed in MW302R, heat-pulse flowmeter profiles were used to calculate the head and the transmissivity of the fractures in the borehole. The results are summarized in table 15.

Continuous water levels were recorded in MW302R to assess any possible effect of pumping from the nearby domestic well on Carriage House Road. A transducer and data logger were installed in MW302R in the fall of 2001. Most of the time, the transducer measured water levels on 15-min intervals. For selected periods of time, water-level data were collected on 1-min intervals.

The continuous record of MW302R showed semi-diurnal fluctuations in the ground-water levels (fig. 22). A small section of the MW302R record is expanded from October 26 to November 4, 2001, which is a time period when there are high demands for ground water. The water-level fluctuations do not coincide with regularly anticipated pumping schedules. Rather these fluctuations are indicative of earth-tide influences caused by the moon and sun's gravitational pull. The cyclic pattern observed in MW302R has a shape similar to the plot of gravimetric forces (fig. 7).

Table 15. Heads and transmissivity derived from heat-pulse flowmeter measurements in MW302R in the UConn landfill study area, Storrs, Connecticut, November 18, 2001.

Depth, in feet below top of casing	Transmissive fracture zone	[1]Relative head, in feet	Water-level elevation, in feet	Transmissivity, in feet squared per day	Proportion of transmissivity, in percent
220.0	1	24.71	552.49	6.8	85
301.0	2	26.81	550.39	1.2	15

[1] Relative head is the magnitude of head for the zone relative to the head of the lowermost zone.

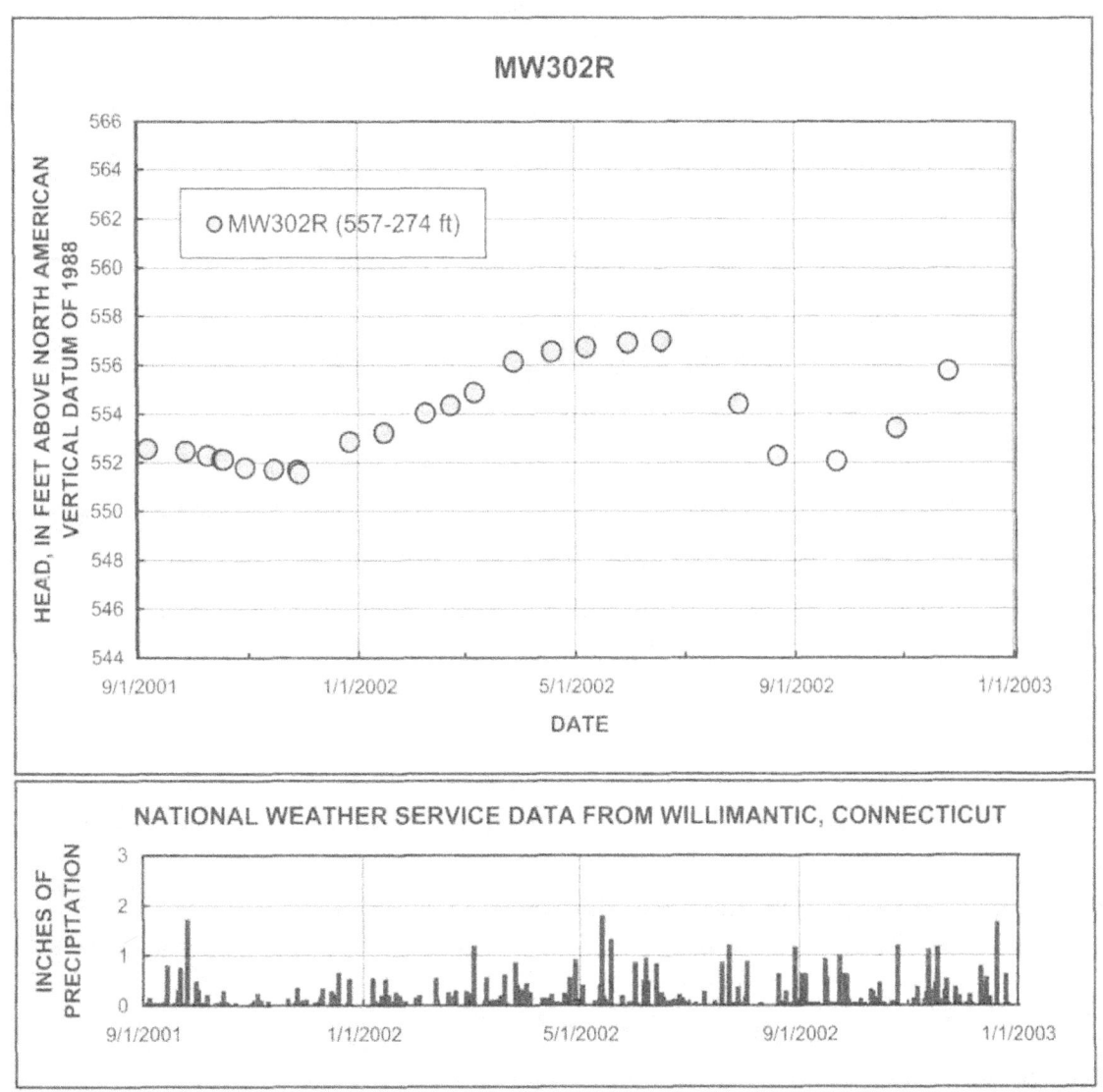

Figure 21. Hydrograph for MW302R in the UConn landfill study area, Storrs, Connecticut. Values in parentheses next to the symbols indicate the elevation of open hole for that monitoring zone. Precipitation and packer-pressure data are shown below the hydrograph. Gaps in the record indicate missing or erroneous data that were omitted.

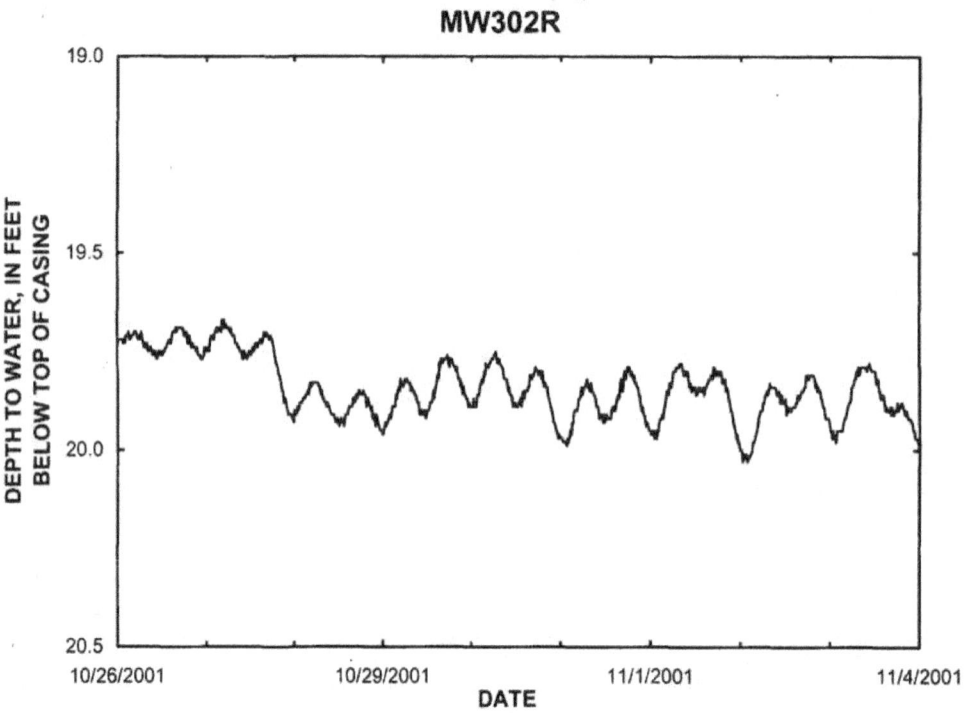

Figure 22. Continuous water-level data for selected period of record for MW302R in the UConn landfill study area, Storrs, Connecticut.

Cross Connections Between Boreholes

Cross-hole testing was done in the area of the former chemical-waste disposal pits to assess possible cross connections between boreholes MW103R, MW104R, MW105R, MW122R, MW203R, MW204R, (piezometer) MW104, MW123SR, and MW303SR. Discrete-interval water-level measurements collected during quarterly water-quality sampling were used to identify drawdown that might indicate the presence of hydraulic connections between boreholes. Additional pumping tests were conducted for the sole purpose of assessing the cross connections in the area of the former chemical-waste disposal pits. The pumping times and monitoring locations are shown in table 4.

The tests entailed pumping from a single zone while monitoring water levels in other zones in the same borehole, in nearby boreholes completed in bedrock, in other shallow bedrock wells, and in overburden wells. No hydraulic impacts were identified in response to pumping in the DZM systems. The lack of hydraulic response indicates a poor connection among fractures in the bedrock between the boreholes that were tested. No change in water level was observed in overburden well MW104, even in response to pumping in the uppermost zone of MW104R. This lack of response indicates that there is no direct hydrologic connection between MW104 and Zones 1 and 5 in MW104R, or that the pumping in Zones 1 and 5 did not subsequently stress the water level in overburden well MW104.

Discussion of Results at the UConn Landfill Study Area

Comparison of heat-pulse flowmeter, DZM system, BAT[3], and packer test methods to determine head and transmissivity. The head values determined with the BAT[3] and the heat-pulse flowmeter profiling with modeling correlated fairly well with the results of the continuous monitoring using the DZM systems. The observed differences in head resulting from these two methods may in part be because the BAT[3] and heat-pulse flowmeter measurements were collected on different days and, therefore, possibly under different hydraulic conditions. BAT[3] and flowmeter methods provide only a snapshot of the hydraulic conditions at a specific time. The heat-pulse flowmeter and BAT[3] methods demonstrated that they are effective screening tools for head determinations. DZM systems, however, provide continuous data for determining head variations under different seasonal conditions.

Transmissivity also was estimated using the BAT[3] and heat-pulse flowmeter data. Only one well that was measured with the BAT[3] was sufficiently transmissive to collect heat-pulse flowmeter measurements that could be modeled with Paillet's method (2000).

The results of heat-pulse flowmeter modeling were compared to the results of packer tests that were reported by Haley and Aldrich, Inc. and others (2002). The straddle-packer apparatus used by Haley and Aldrich, Inc. did not measure the head of zones above and below the test zone. Hence, comparison between measured and modeled heads cannot be made. A comparison of transmissivity estimates made with the two methods is shown in figure 23A. For all but one borehole, the transmissivity estimated from heat-pulse flowmeter results and transmissivity estimated from packer tests were within an order of magnitude. The flowmeter- and packer-test-derived transmissivity values were converted to log values and plotted at a linear scale for correlation (fig. 23B). The correlation coefficient using all log transmissivity values was 0.461. Two of the packer tests each had two estimates of transmissivity – one for the early-time data (less than 10 min) and another for later-time data. The later-time packer-test results probably are more comparable to the flowmeter results, because the flowmeter results are based on quasi steady-state conditions achieved after many minutes of pumping. By removing the two outlier transmissivity results that were fit to early-time data, the correlation coefficient improved to 0.910. These results indicate that the heat-pulse flowmeter may be sufficient for measuring transmissivity to within an order of magnitude. Moreover, the heat pulse flow meter data can be collected in less time than with BAT[3] or packer tests in open holes. If discrete-interval samples need to be collected, however, or if the transmissivity is lower than the measurement range of the flowmeter, then discrete-interval packer testing is preferred.

Open-hole head compared to discrete heads. A comparison of open-hole and discrete-interval heads illustrates the concept that open-hole head is a transmissivity-weighted average of the individual heads. A good example is in MW105R, where the open hole is dominated by the water level of the deepest zone, which was identified as the most transmissive. This concept was useful in assessing the possibility of packer failure, as the heads tended to converge on the head of the most transmissive zone.

Trends in water levels. In five of the boreholes, water levels in the DZM systems were recorded for 2 yr, and water levels in the other four boreholes were monitored for 1.5 yr. Although this period of record is short for assessing long-term change, there are no obvious long-term linear trends in the data. A water-level decline occurs seasonally, but the boreholes appear to recover fully. A longer period of record would be needed to assess long-term trends more rigorously.

Daily fluctuations in water levels. The amplitude and periodicity of short-term fluctuations in water levels were analyzed for the effects of earth tides. Small semi-diurnal water-level fluctuations in this study were determined to be a result of earth-tide influences. No fluctuations typical of pumping were identified in the water-level record.

Seasonal fluctuations in water levels. The hydrographs show that the water levels fluctuate with seasonal changes in precipitation, recharge, and evapotranspiration. Head in discrete zones and in different boreholes varies both in magnitude of response and in timing of response to precipitation. In general, there were downward driving potentials in the recharge areas and in the area of the ground-water divide, and upward

driving potentials in discharge areas north and south of the landfill. Moreover, the water levels in recharge areas showed large changes in response to drought. Discharge areas showed less change than the recharge areas – there were only subtle changes in the heads in boreholes in ground-water discharge zones (MW101R, MW105R, and MW201R) relative to the changes in head in recharge zones (MW109R, MW202R, and MW122R). This is illustrated by a comparison of the range of water-level fluctuations from spring to fall in MW109R and in MW101R – the most dramatic change in head (about 17.5 ft) was observed in MW109R in a recharge zone, and the most subdued change (about 2 ft) occurred in MW101R in a discharge zone.

Comparison of hydrographs along conceptual flow-paths in the study area. Although water levels fluctuate and the magnitude of the driving potentials between fractures changes, the direction of flow typically is the same during dry and wet periods. The direction of vertical driving potentials within each borehole generally was consistent throughout the period of investigation. Exceptions to this include (1) MW201R and MW105R, where there were seasonal reversals in the driving potentials within each borehole; and (2) MW122R, MW204R, and MW121R, where the heads show downward driving potentials during periods of recharge and vertical equi-potentials in the absence of recharge.

In addition, the direction of driving potential between the boreholes remained fairly consistent. A composite hydrograph constructed for boreholes MW204R, MW103R, MW101R, and piezometer MW101 shows progressively lower heads with distance from the ground-water divide north toward the wetland (fig. 24). The hydrograph shows that the direction of driving potentials from one borehole to another is consistent over the entire period of record. The water levels in all zones in MW204R are higher than water levels in all zones in MW103R, and water levels in all zones in MW103R are higher than in all zones in MW101R. This profile indicates the potential for a northern flow direction in this region of the study area.

Similarly, a composite hydrograph was constructed for boreholes that are located along a profile from the ground-water divide (north of the chemical-waste disposal pits) and extending south away from the landfill area (fig. 25). The hydrograph shows progressively lower heads from the north to the south. The heads in the boreholes in the area of the chemical pits, including MW203R and MW104R, are consistently higher than the heads in MW105R, which are always higher than the heads in MW201R. This head distribution shows a driving potential towards the south and indicates the presence of a southern flow path that extends from the former chemical-waste disposal pits to beyond MW201R. In addition, upward driving potentials occur seasonally at MW105R and year round at MW201R. There also are upward driving potentials between MW201R and piezometer 13, which is completed in the shallow bedrock, and in the wetlands south of piezometer 13.

A composite hydrograph constructed for boreholes along an east-west profile shows that the heads for MW104R, which is west of the landfill and east of the former chemical-waste disposal pits, are among the lowest along the profile (fig. 26).

(A)

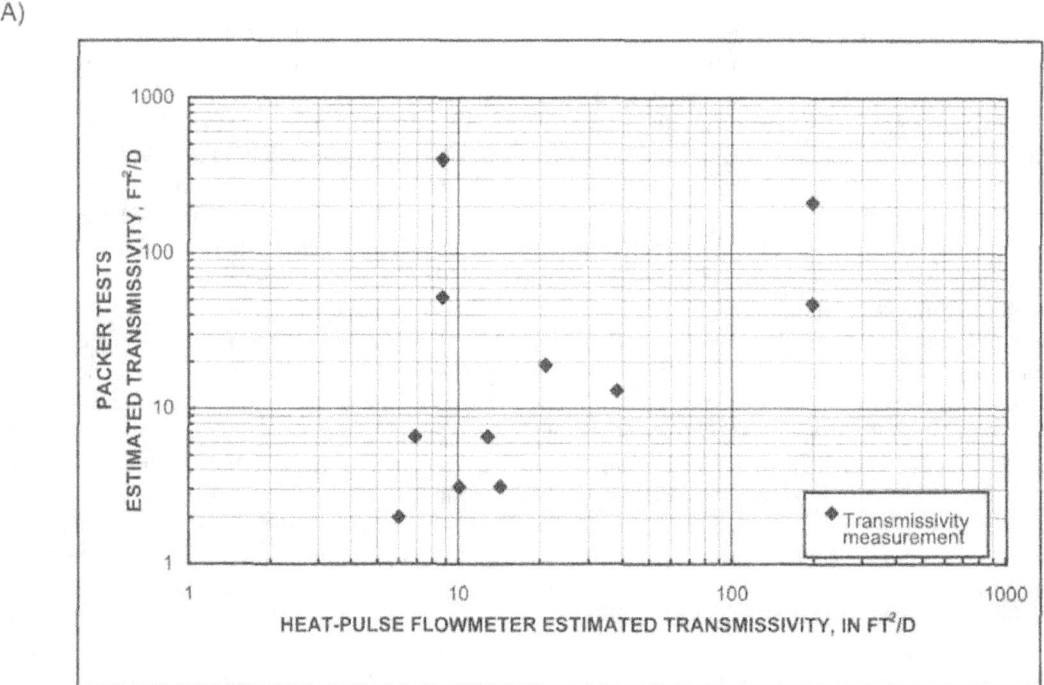

(B)

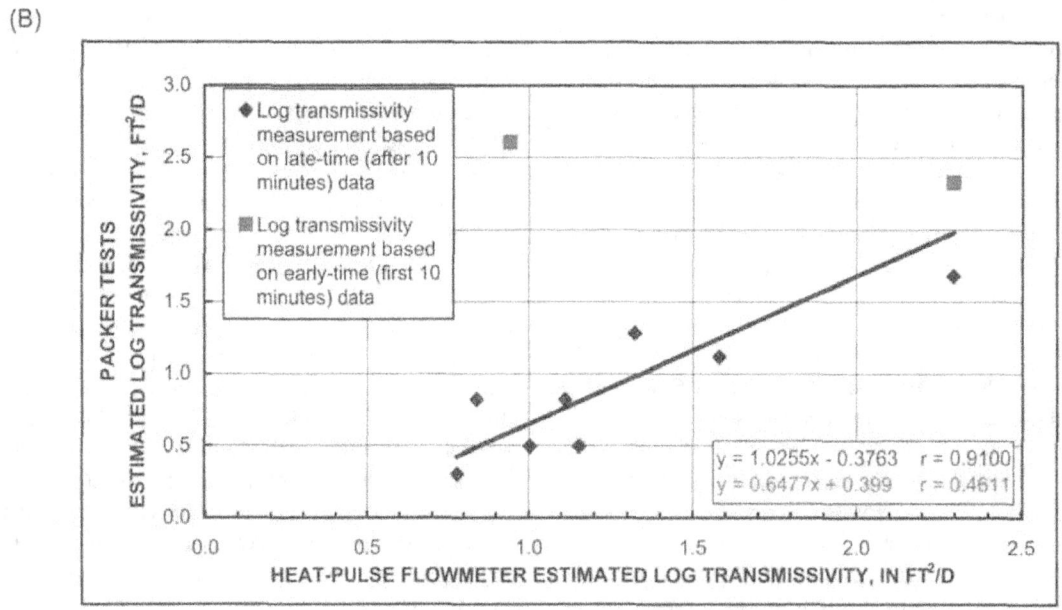

Figure 23. (A) Comparison of open-hole transmissivity estimates in feet-squared per day (ft²/d) from the heat-pulse flowmeter and packer tests in selected boreholes in the UConn landfill study area, Storrs, Connecticut. (B) Comparison of the log values of open-hole transmissivity estimates from the heat-pulse flowmeter and packer tests in selected boreholes in the UConn landfill study area, Storrs, Connecticut. Correlation coefficient, r, in red includes all of the transmissivity measurement estimates (red and blue symbols). The correlation coefficient, r, in blue only includes estimates based on late-time packer-test data (blue symbols only). Blue trendline is for late-time data.

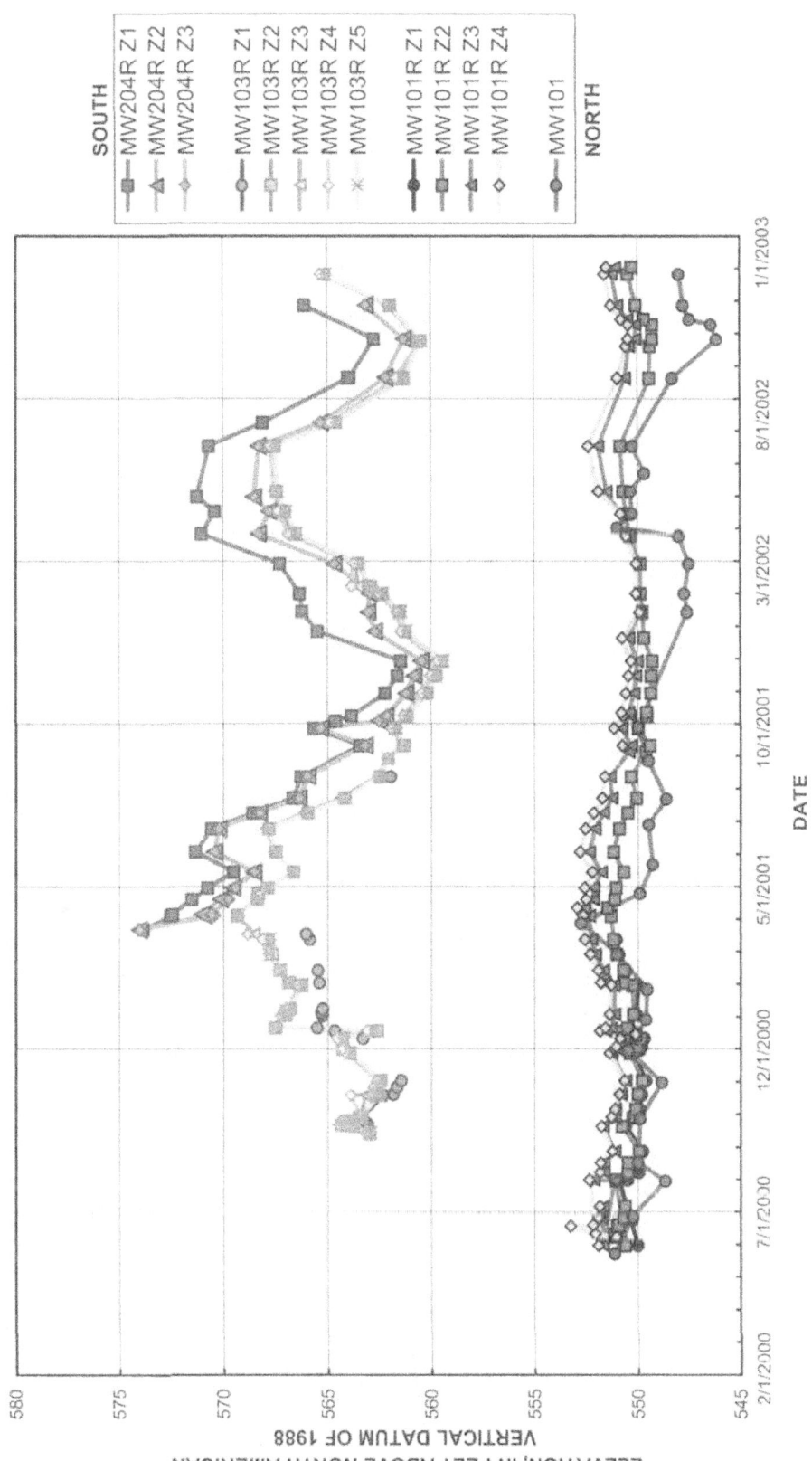

Figure 24. Composite hydrograph for discretely monitored zones in bedrock boreholes and piezometers along the northern flowpath of the UConn landfill study area, Storrs, Connecticut. Gaps in the record indicate missing or erroneous data that were omitted. Hydrograph indicates the heads are progressively lower to the north along flowpath. Within a single borehole, heads are lower in the deeper zones in recharge locations. Heads are lowest at the shallowest zones in discharge locations, indicating upward driving potentia for flow.

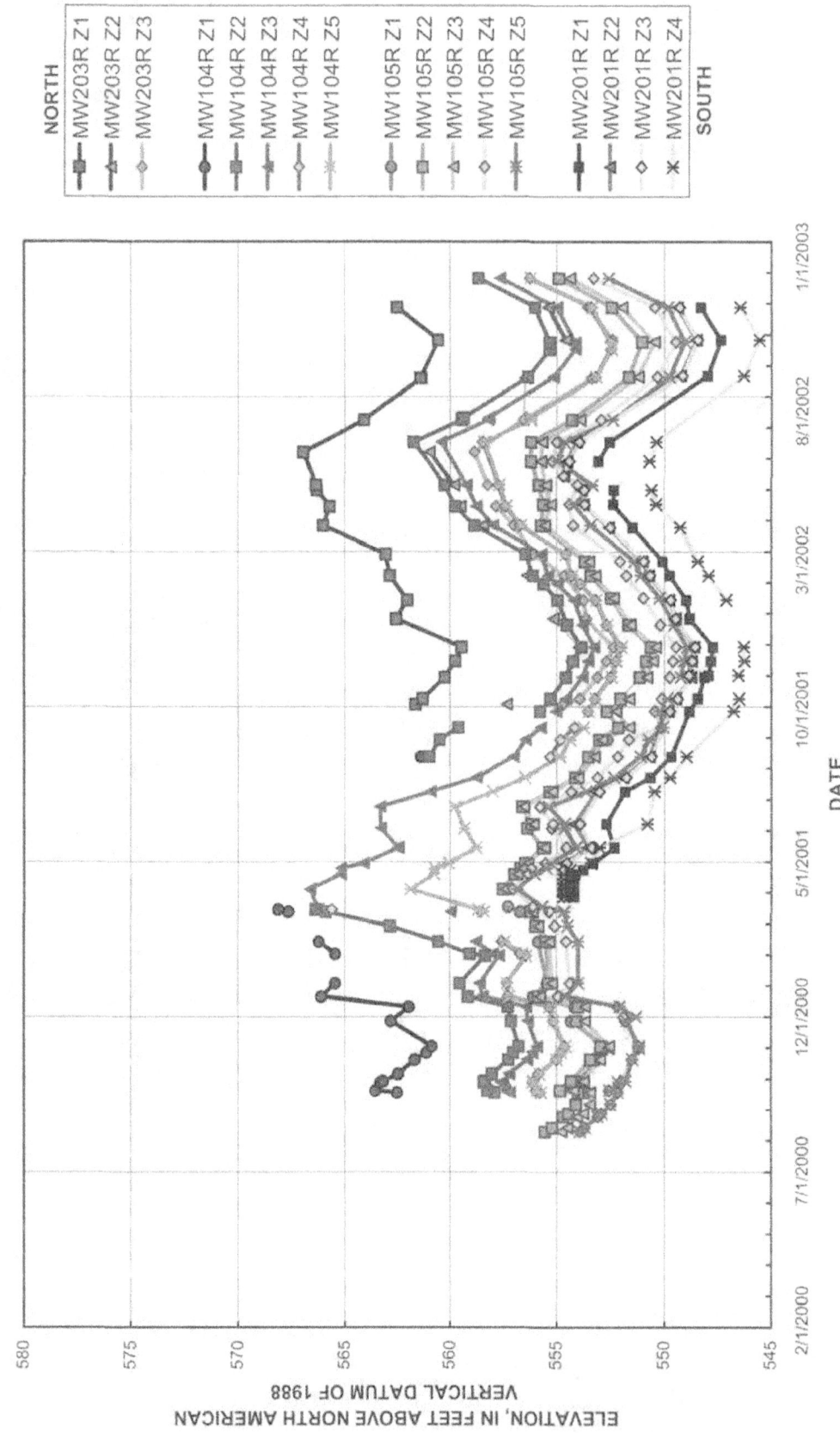

Figure 25. Composite hydrograph for discretely monitored zones in bedrock boreholes and piezometers along the southern flowpath of the UConn landfill study area, Storrs, Connecticut. Gaps in the record indicate missing or erroneous data that were omitted. Hydrograph indicates heads are progressively lower to the south. Within a single borehole, heads are lower in the deeper zones in recharge locations. Heads in MW201R indicate some upward gradients, specifying upward driving potentials in the discharge location.

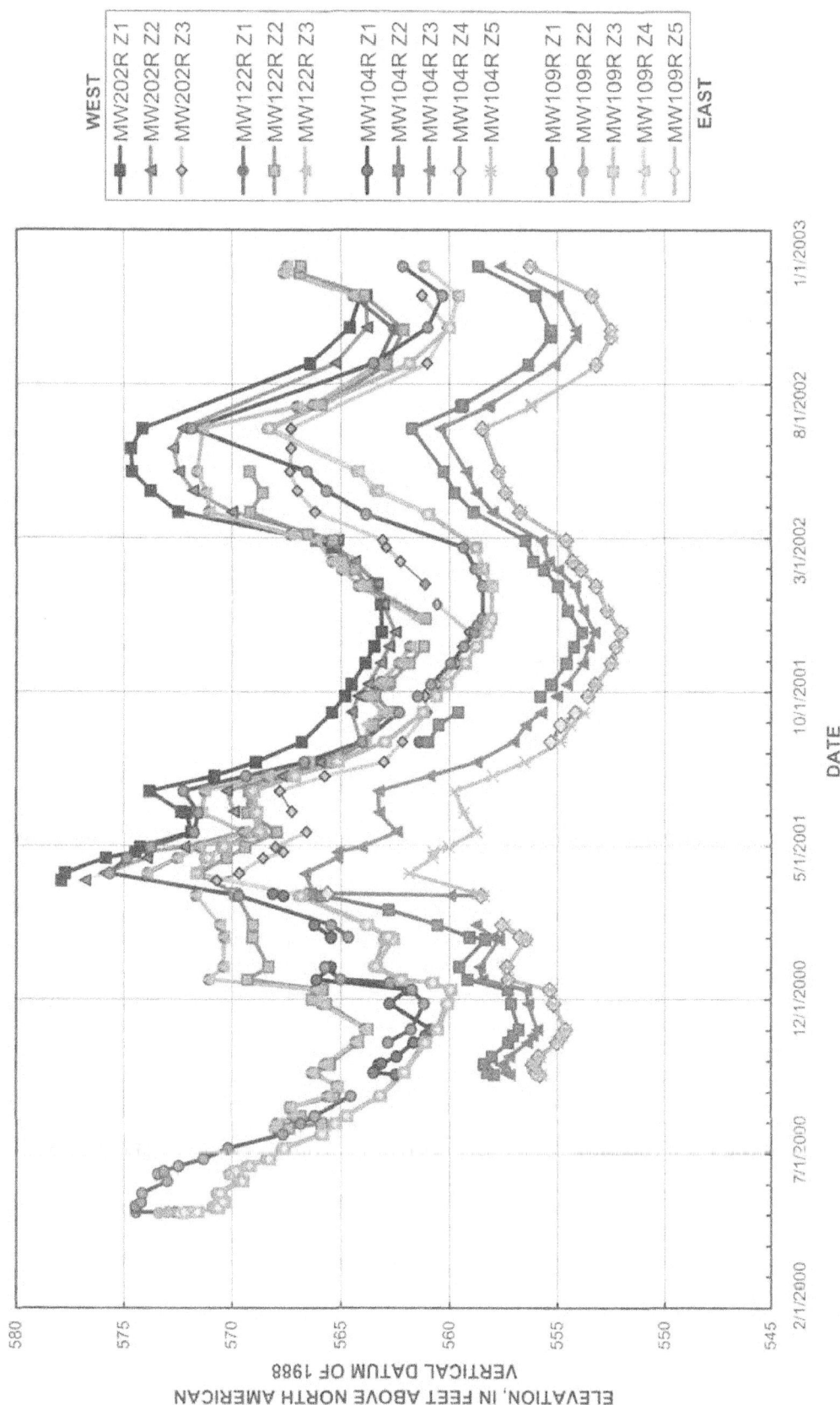

Figure 26. Composite hydrograph for discretely monitored zones in bedrock boreholes and piezometers along the west-east flowpath of the UConn landfill study area, Storrs, Connecticut. Gaps in the record indicate missing or erroneous data that were omitted. Hydrograph indicates the lowest heads are in the middle of the valley at MW104R and the heads east and west of the landfill are higher than at MW104R. Within a single borehole, heads generally decrease with depth indicating these are recharge zones.

Heads in the boreholes on the hills to the east (MW109R) and the hills to the west (MW122R and MW202R) are consistently higher than water levels in the valley. The driving potentials in each of the boreholes along the profile are downward, which is consistent with boreholes placed in recharge zones. The hydrograph also shows that the heads in zones in MW202R generally are higher than the heads in zones in MW122R, which are higher than the heads in MW104R. These results indicate that there is driving potential from the hill slopes towards the valley, and this pattern is consistent over the depths of investigation in the DZM systems.

Verification of the conceptual ground-water flow model. The pattern of hydraulic head and driving potentials is consistent with the conceptual model for ground-water flow. Specifically, the measured hydraulic potentials between the overburden and the fractured-bedrock aquifers indicate these aquifers are hydraulically connected. Hydrographs indicate that water discharges from the overburden and recharges the bedrock in the topographically higher areas in the study area, and water discharges from the bedrock into the overburden in the topographically lower areas. Water-quality data collected from the bedrock wells also support this conceptual model, with the highest concentrations of contaminants in the source areas and a distribution of contaminants in a north-south direction along the axis of the valley (Johnson and Kastrinos, 2002).

Cross sections were constructed along the axis of the north-south valley (A-A' and B-B') and in the east-west direction from the topographic high on the east of the landfill to Hunting Lodge Road west of the landfill (C-C') to compare the hydraulic head distributions and water-quality data (fig. 1). The distribution of hydraulic heads along the profiles was compared to one-time water-quality samples collected from ground-water profiling points in 1999 and to water-quality samples obtained quarterly from 1999 through 2002 from piezometers in the overburden, discrete zones in boreholes completed in bedrock, and open-hole domestic wells completed in the bedrock.

After about four rounds of sampling, characteristic chemical signatures were established for ground water impacted by landfill leachate and by the contaminants from the former chemical-waste disposal pits (Haley and Aldrich, Inc. and others, 2002). Elevated specific conductance, high iron and cadmium, negative oxidation-reduction potential, and chlorobenzene are collectively indicative of the presence of landfill leachate. The chemical signature from the former chemical-waste disposal pits includes low specific conductance, positive oxidation-reduction potential, and the presence of chlorobenzene and chlorinated compounds such as tetrachloroethene, trichloroethane, and trichloromethane.

The results of selected water-quality samples were superimposed on the cross sections. Target volatile organic compounds (VOCs), which include benzene, chlorobenzene, ethylbenzene, tetrachloroethene, toluene, trichloroethane, and xylenes, were used as an indicator of contamination from the former chemical-waste disposal pits and from the landfill. Maximum concentrations of target VOCs collected from 1999 through 2002 are shown on the profiles A-A', B-B', and C-C'

with solid circles for samples collected in bedrock and open circles for samples collected from the overburden (figs. 27B, 28B, and 29B, respectively). The highest concentrations are shown in hot colors (red and orange), and the lowest concentrations are shown in cool colors (blue and black). Samples that did not have detections of target VOCs are shown with a plus sign (+).

Other chemical constituents collected at synoptic water-quality sampling events also were evaluated to verify the conceptual ground-water flow model for the site. Only the results of the maximum target VOCs are shown in this summary. The results of all chemical sampling are presented and discussed in detail by Haley and Aldrich, Inc. and others (2002). The examples provided here illustrate the combined use of discrete-interval water-quality and hydraulic-head data from isolated zones to verify a ground-water flow conceptual model. The observed distribution of target VOCs along the profiles is consistent with the pattern of hydraulic heads and driving potentials observed in the DZM systems and in piezometers along the profiles and thus supports the conceptual model of ground-water flow in the study area.

The southern flowpath. The cross section A-A' (fig. 1) extends from the ground-water divide south along the axis of the valley beneath the former chemical-waste disposal pits. Green arrows in figure 27A show the direction of driving potential measured in the DZM systems in the boreholes along the profile. The dashed up and down arrows at the domestic well W202-NE indicate the vertical driving potentials that were measured with the heat-pulse flowmeter (on April 26, 2001). Blue arrows show the generalized flow directions between boreholes.

The maximum occurrence of target VOCs that were collected from discrete intervals along the profile from 1999 through 2002 is shown in figure 27B (John Kastrinos, Haley and Aldrich, Inc., written commun., 2003). The distribution of target VOCs is consistent with the pattern of hydraulic heads and driving potentials observed in the DZM systems along the profile. Although the contaminated sediments in the overburden were removed in 1987, some of the contamination leached into the bedrock, as indicated by the observation of the highest concentrations of VOCs (shown in red and orange) in the shallow bedrock (MW303SR, MW123SR, and MW203R, fig. 27B) near the former chemical-waste disposal pits. Near the ground-water divide and in the area of the former chemical-waste disposal pits, the downward driving potentials spread the target VOCs downwards (MW203R and MW104R, fig. 27A) from the overburden and into the bedrock. Target VOCs were observed at depth in borehole MW104R. Farther downgradient at MW105R, target VOCs were observed at the sampling intervals at depths of 74 and 110 ft below the top of the casing; however, the maximum concentrations were at least an order of magnitude lower than the maximum concentrations that were observed at the source area. These results are consistent with the hydraulic heads and driving potentials observed in MW105R and consistent with the flowpaths inferred by the hydraulic head observed between MW104R and MW105R.

Farther downgradient along the profile at MW201R, the VOCs were observed at the sampling intervals at 60 and 38 ft. The chemical distribution is consistent with the interpretation that this is a discharge location, and the chemicals are discharging upwards. Concentrations of target VOCs in piezometer 13 and in the overburden samples farther south of MW201R were less than 1 µg/L or nondetectable. In well W202-NE, an open-hole sample, which represents a transmissivity-weighted mixture of water from all productive fractures intersecting W202-NE, showed a maximum target VOC concentration of 4.7 µg/L in September 2000. A discrete-interval sample from a depth of 197 ft also indicates the presence of VOCs (0.7 µg/L) in W202-NE, indicating the presence of a deep bedrock migration pathway with chemical constituents near the analytical detection limit.

The northern flowpath. The cross section B-B' (fig. 1) extends from the ground-water divide on the west side of the landfill to the northern end of the landfill (fig. 1). For each borehole along the profile, the directions of measured driving potentials are indicated with green lines and arrows in figure 28A. Downward potentials were observed in MW204R near the ground-water divide. Upward driving potentials were observed between the bedrock and overburden and in the bedrock at boreholes west and north of the landfill (MW103R and MW101R). A comparison of the hydraulic heads in each of the boreholes along the profile in figure 28A indicates the potential for flow toward the north. The direction of potential flow between boreholes is depicted with blue lines.

Water-quality samples collected from overburden and shallow bedrock north of the landfill indicate the presence of landfill leachate (Haley and Aldrich, Inc. and others, 2002). Electromagnetic terrain-conductivity surveys collected in 1998, 1999, and 2000 north of the landfill indicated the magnitude of apparent conductivity decreased with depth and with distance from the landfill (Powers and others, 1999; Johnson and others, 2002). The decrease in apparent conductivity was interpreted as the result of decreasing concentrations of landfill leachate in ground water.

The highest VOC concentrations along the northern flowpath were observed in the piezometers in the overburden near the base of the landfill in an area where there are upward potentials (fig. 28B). These target VOCs observed in piezometers MW101 and MW103, at ground-water profiling point GW4, and in borehole MW103R were interpreted as coming from localized VOC sources in an area of downward driving potential within the landfill (Haley and Aldrich, Inc. and others, 2002), rather than originating from a source at the former chemical-waste disposal pits. The lower concentrations and nondetections were observed in the downgradient piezometers located in the wetland north of the landfill. This VOC concentration distribution is consistent with the apparent conductivity data. Similar analyses conducted for the compounds indicative of landfill leachate indicate that leachate discharges upward to the wetland and is dissipated in the wetland within 500 ft of the landfill (Haley and Aldrich, Inc. and others, 2002).

The east-west profile, C-C', extends from Hunting Lodge Road east across the ground-water divide and the landfill up the slope of the hill to the east of the landfill (fig. 1). The direction of driving potential observed in the DZM systems is indicated for each borehole in figure 29A. The distribution of vertical heads along profile C-C' indicates strong downward driving potentials on the slope of the hill east of the landfill (MW109R and MW109); seasonal downward driving potentials in the area of the former chemical-waste disposal pits; and seasonal downward driving potentials west of the landfill. The heads in the bottom zones of MW109R were slightly higher than the head in the middle zone, indicating the potential for upward flow. This pattern of downward driving potentials in the upper part of MW109R and upward driving potentials in the lower part of MW109R is consistent with the localized flowpaths on the side of a hill slope (Winter and others, 1998).

The driving potentials along the profile C-C' are less than the driving potentials measured along the axis of the valley, as indicated in profiles A-A' and B-B' (figs. 27A and 28A, respectively). Comparison of heads between the boreholes along profile C-C' indicates potential for flow towards (MW104R) near the center of the valley. Blue arrows show the generalized flow directions between the boreholes that are shown along profile C-C'.

During periods of recharge, there are downward driving potentials near the ground-water divide and there are driving potentials between wells in the east-west direction (fig. 29A) and in the north-south direction (figs. 27A and 28A). When there is no precipitation and no recharge, the hydraulic heads from discrete intervals in individual boreholes indicate vertical equipotentials. For example, a comparison between the water levels in the discrete zones in MW122R showed 2 ft of differential head during recharge and 0-ft difference during drought. These results indicate that during drought periods, there are not strong downward driving potentials. The hydraulic heads in MW104R, however, are consistently lower than the heads in the boreholes to the east (MW109R) and to the west (MW122R and MW202R), indicating that the local topographic highs cause potential for flow towards the center of the valley. A comparison of hydraulic heads between the boreholes along the axis of the valley indicate that the hydraulic heads in the boreholes in the area of the former chemical-waste disposal pits are consistently higher than the heads in the boreholes to the south and to the north.

Maximum target VOCs that were collected quarterly over the period of investigation (1999 through 2002) are shown on the profile in figure 29B. The highest VOC concentrations were observed in the piezometers and shallow bedrock near the former chemical-waste disposal pits. To the east and west of the former chemical-waste disposal pits, the target VOCs were orders of magnitude lower than in the area of the former disposal pits.

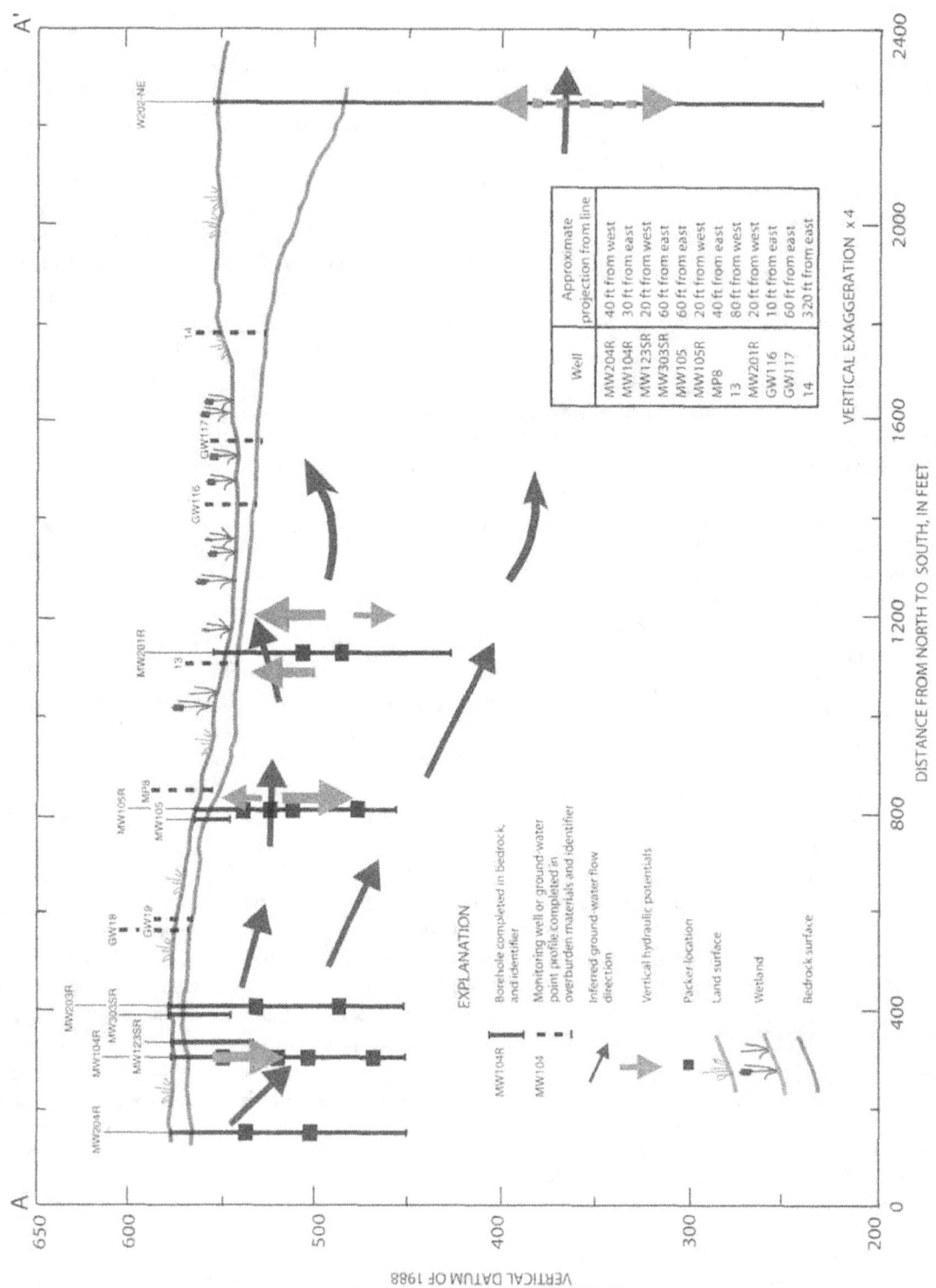

Figure 27A. Flow paths and vertical hydraulic potentials in the southern part of the UConn landfill study area, Storrs, Connecticut along line A-A' depicted in figure 1. The green arrows indicate the direction of hydraulic potential measured in the discrete-zone monitoring systems. The dashed green line indicates hydraulic potentials determined from heat-pulse flowmeter measurements. Blue arrows show the inferred ground-water flow direction. Approximate projection magnitudes and directions are listed because boreholes, monitoring wells, and ground-water profiling points not located on line A-A' are projected onto the line from their actual locations.

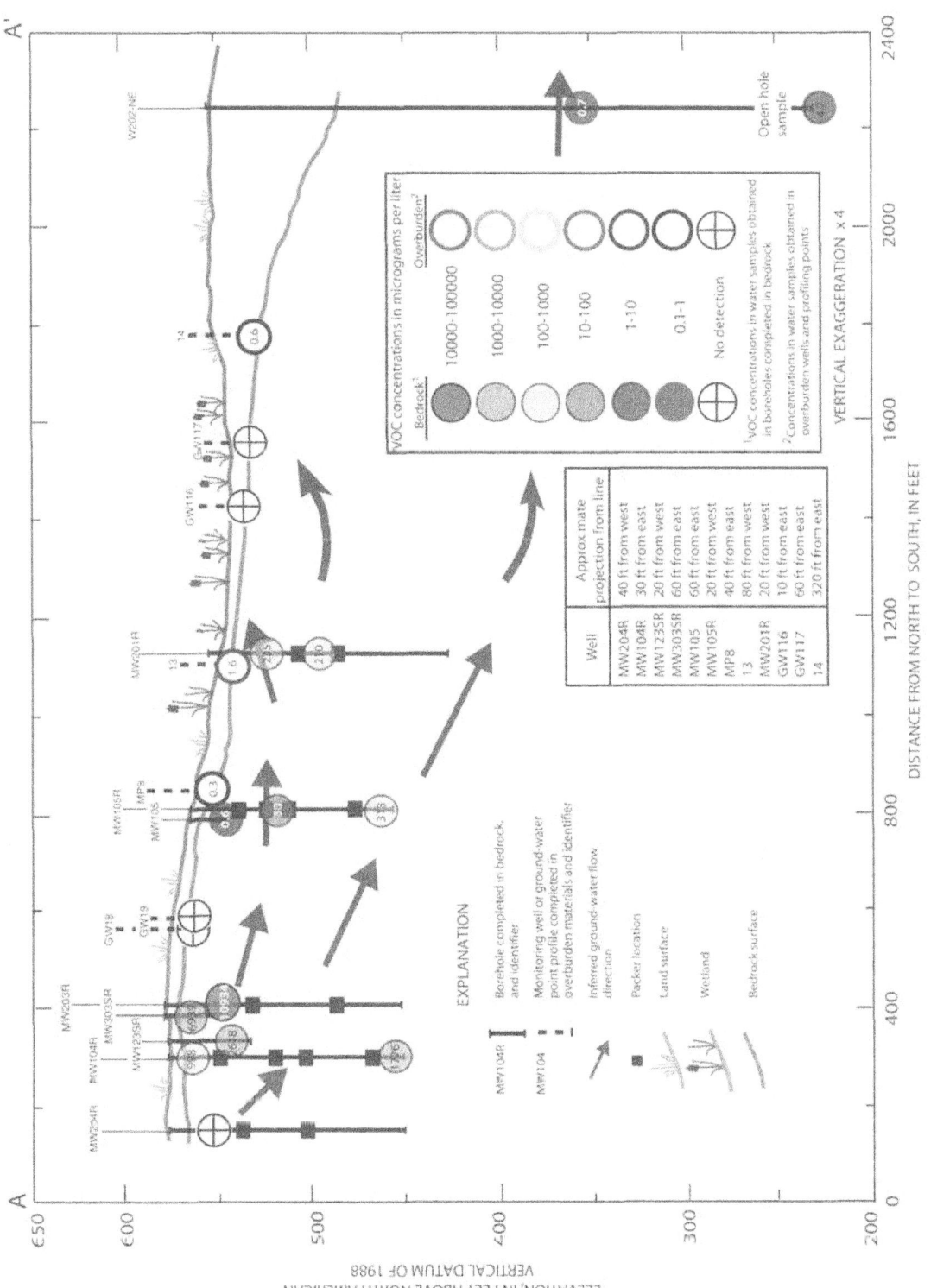

Figure 27B. Maximum concentrations of target volatile organic compounds (VOC) (including benzene, chlorobenzene, ethylbenzene, tetrachloroethene, toluene, trichloroethane, and xylenes) observed in water samples collected from 1999 through 2002 along line A-A' depicted in figure 1. Solid-colored circles indicate VOC concentrations observed in bedrock samples and ground-water profiling points, and open circles indicate overburden samples. The color of the circle indicates the range of VOC concentration and the actual concentration is shown inside the circle in micrograms per liter. Blue arrows show the inferred ground-water flow direction. Approximate projection magnitudes and directions are listed because wells not located on line A-A' are projected onto the line from their actual locations.

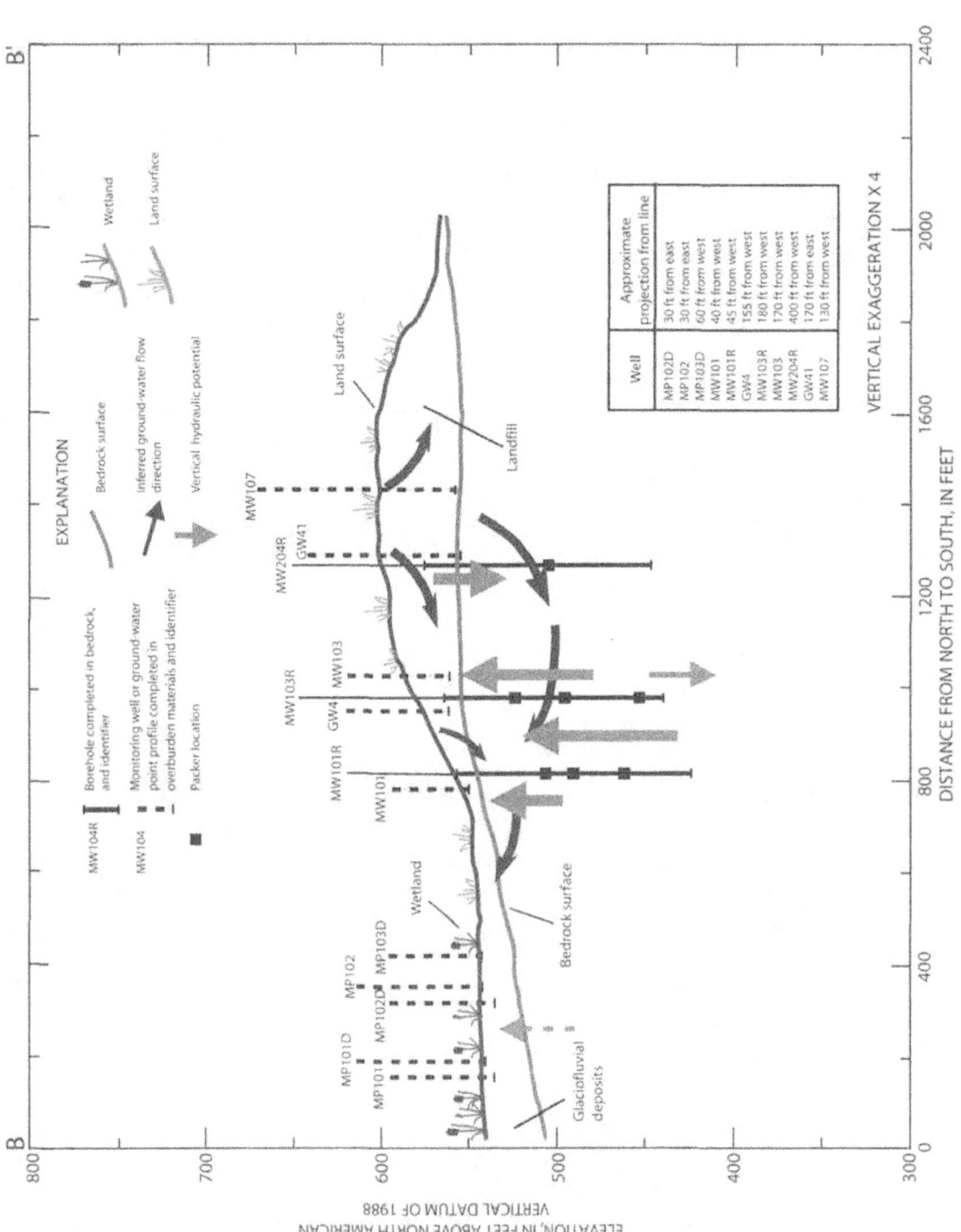

Figure 28A. Flow paths and vertical hydraulic potentials in the northern part of the UConn landfill study area, Storrs, Connecticut along line B-B' depicted in figure 1. The green arrows indicate the direction of hydraulic potential measured in the discrete-zone monitoring systems. The dashed green line indicates hydraulic potentials determined from heat-pulse flowmeter measurements. Blue arrows show the inferred ground-water flow direction. Approximate projection magnitudes and directions are listed because boreholes, monitoring wells, and ground-water profiling points not located on line B-B' are projected onto the line from their actual locations.

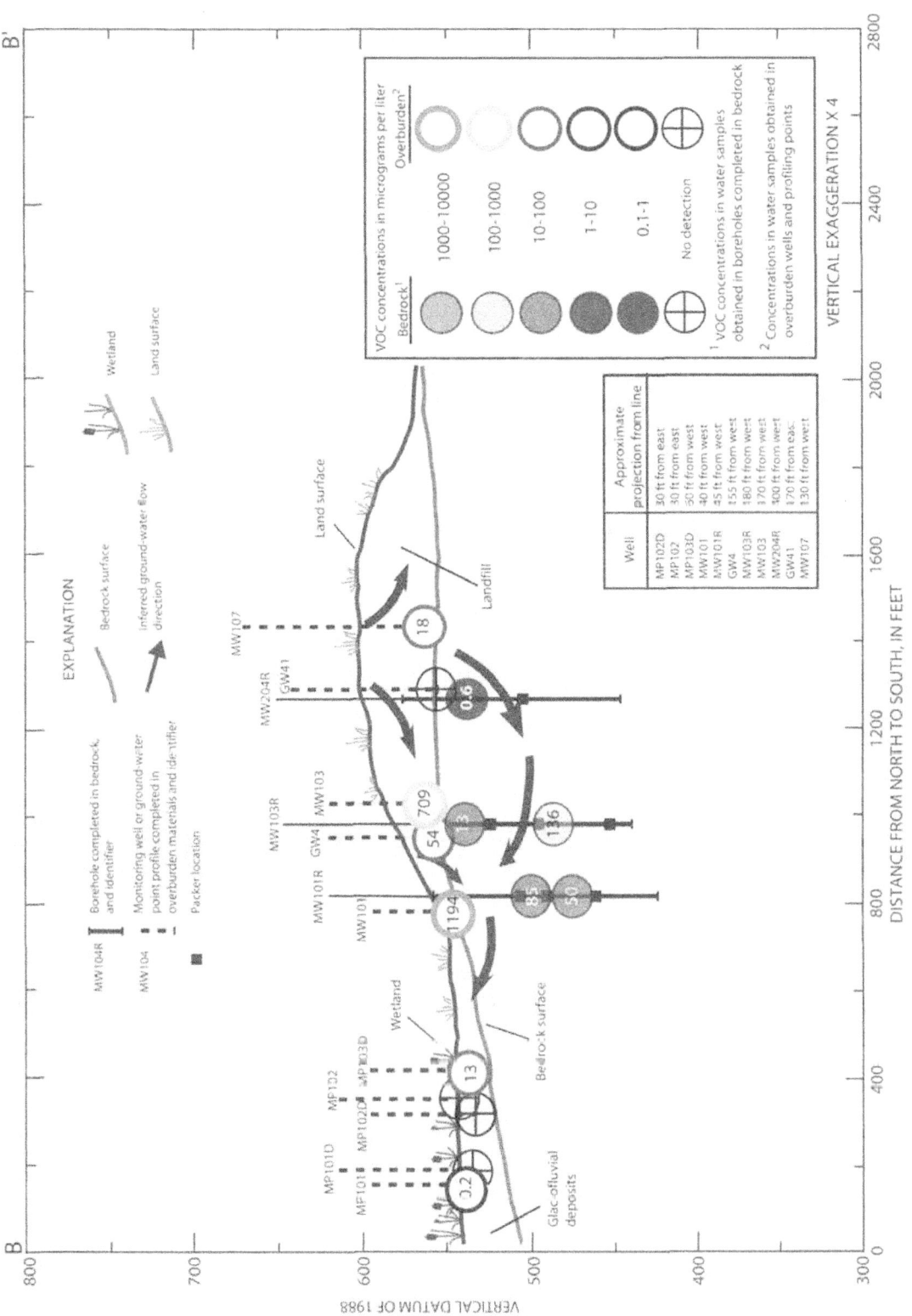

Figure 28B. Maximum concentrations of target volatile organic compounds (VOC) (including benzene, chlorobenzene, ethylbenzene, tetrachloroethene, toluene, trichloroethane, and xylenes) observed in water samples collected from 1999 through 2002 along line B-B' depicted in figure 1. Solid-colored circles indicate VOC concentrations observed in bedrock samples and ground-water profiling points, and open circles indicate overburden samples. The color of the circle indicates the range of VOC concentration and the actual concentration is shown inside the circle in micrograms per liter. Blue arrows show the inferred ground-water flow direction. Approximate projection magnitudes and directions are listed because wells not located on line B-B' are projected onto the line from their actual locations.

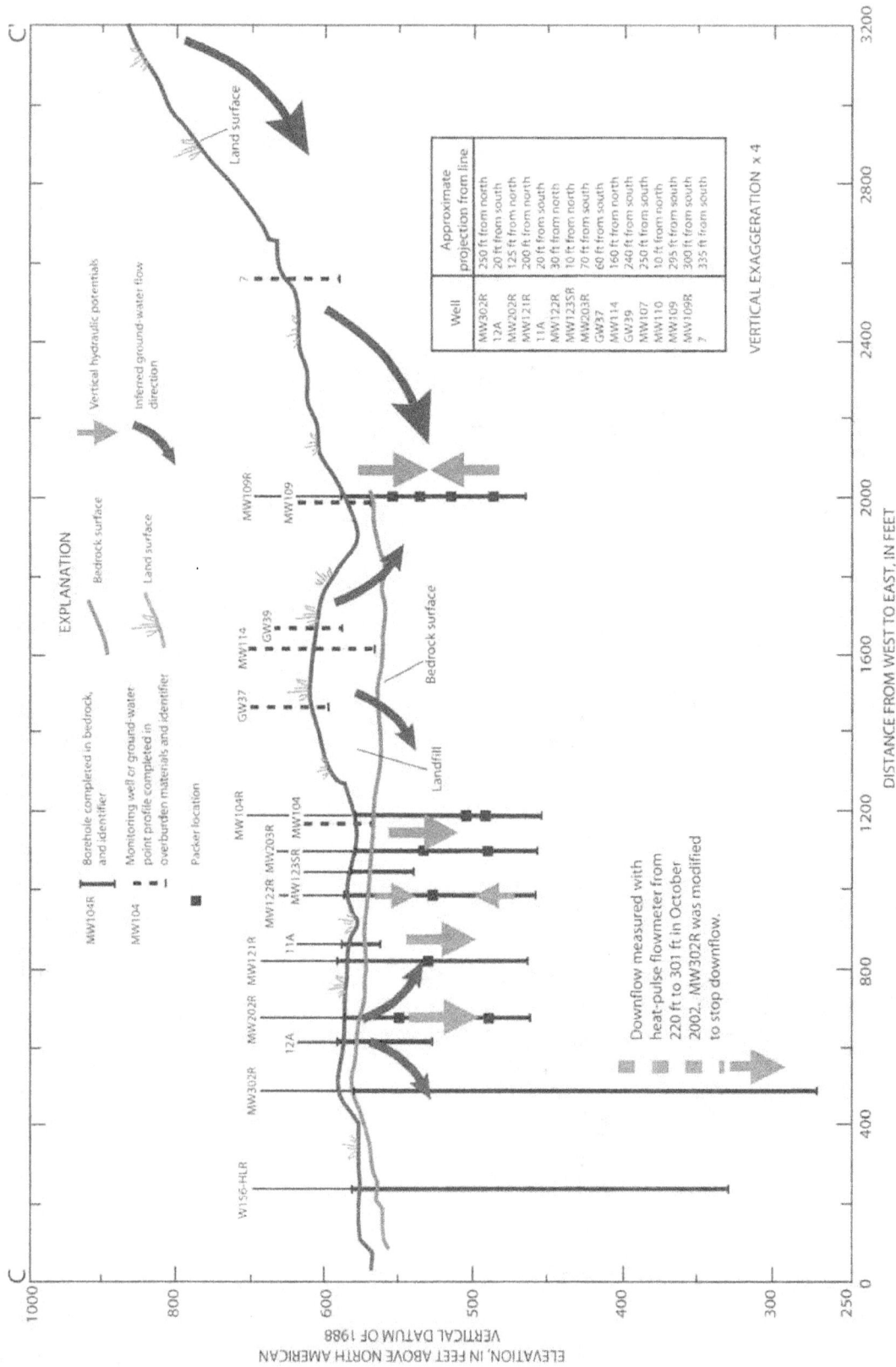

Figure 29A. Flow paths and vertical hydraulic potentials going west to east across the UConn landfill study area, Storrs, Connecticut along line C-C' depicted in figure 1. The green arrows indicate the direction of hydraulic potential measured in the discrete-zone monitoring systems. The dashed green line indicates hydraulic potentials determined from heat-pulse flowmeter measurements. Blue arrows show the inferred ground-water flow direction. Approximate projection magnitudes and directions are listed because boreholes, monitoring wells, and ground-water profiling points not located on line C-C' are projected onto the line from their actual locations.

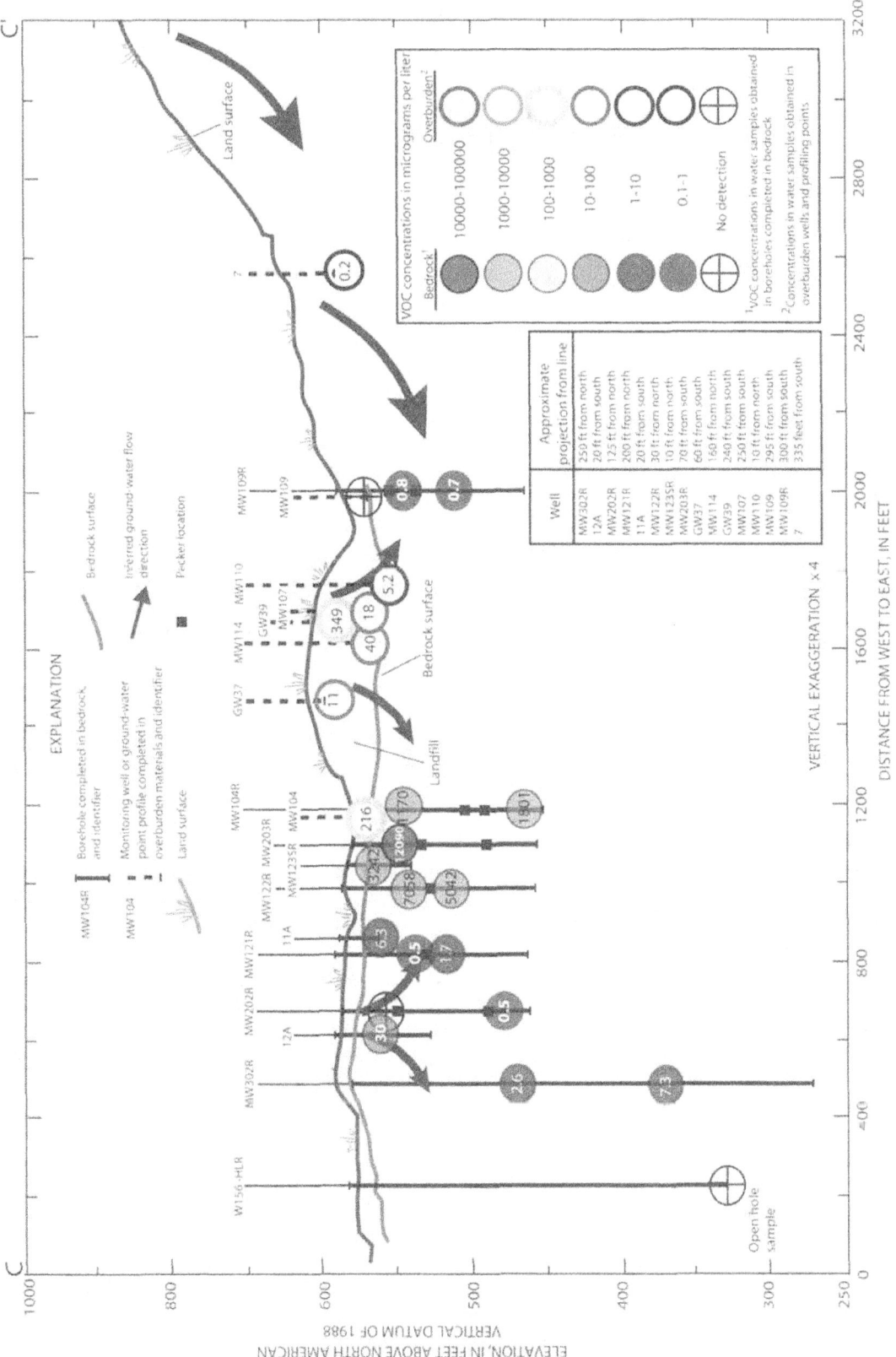

Figure 29B. Maximum concentrations of target volatile organic compounds (VOC) (including benzene, chlorobenzene, ethylbenzene, tetrachloroethene, toluene, trichloroethane, and xylenes) observed in water samples collected from 1999 through 2002 along line C-C' depicted in figure 1. Solid-colored circles indicate VOC concentrations observed in bedrock samples and ground-water profiling points, and open circles indicate overburden samples. The color of the circle indicates the range of VOC concentration and the actual concentration is shown inside the circle in micrograms per liter. Blue arrows show the inferred ground-water flow direction. Approximate projection magnitudes and directions are listed because wells not located on line C-C' are projected onto the line from their actual locations.

The combined use of head distributions from the DZM systems, measurements with the heat-pulse flowmeter, and results of discrete-interval water-quality sampling (target VOCs and other selected parameters such as benzene) has helped refine and verify the conceptual ground-water flow model for the site. Although the open-hole head measurements define the general flow directions, alone they would not have provided the detailed hydraulic-head information needed to determine the vertical head differences and driving potentials along the major flowpaths. Installation of the DZM systems prevented vertical flow through the open boreholes, thereby minimizing the spreading of contaminants and dilution of the water-quality in discrete zones.

Summary and Conclusions

Heat-pulse flowmeter profiles and pumping records, manual open-hole water-level and discrete-zone water-level measurements, continuous discrete-zone water-level measurements, and differential head testing using a straddle-packer apparatus were used to determine hydraulic head, driving potential, and transmissivity in boreholes and piezometers in a fractured-rock aquifer near the former landfill and chemical-waste disposal pits at University of Connecticut (UConn), Storrs, Connecticut. These data were analyzed to identify and characterize relations between long-term water-level patterns and precipitation, topographic setting, contaminant distribution at the site, and the conceptual ground-water flow model at the study area. The analysis helped to establish, refine, and verify the conceptual model of ground-water flow in the study area in order to better determine and explain the distribution of contamination at the site.

The head values determined with the bedrock-aquifer transportable testing tool (BAT[3]) and the heat-pulse flowmeter data correlated fairly well with the results of the continuous monitoring using the discrete-zone monitoring (DZM) systems. These results indicate that the heat-pulse flowmeter and BAT[3] are effective screening tools for head determination. BAT[3] and flowmeter methods provide only a snapshot of the hydraulic conditions at a specific time, whereas the DZM systems provide continuous long-term data for determining head variations under seasonal conditions.

Log transmissivity values determined from heat-pulse flowmeter modeling were regressed against log transmissivity values determined by conventional straddle-packer testing. The regression yielded a good coefficient of determination (R^2 = 0.8282) indicating that the heat-pulse flowmeter can be a reliable method for estimating fracture transmissivity for the most transmissive fractures in a borehole. The strength of the heat-pulse flowmeter method is that data can be collected relatively easily and quickly compared to data collection using the BAT[3] equipment. The flowmeter is not as sensitive, however, as the BAT[3] equipment for determination of head or transmissivity. Because of flowmeter sensitivity, small-yielding (low transmis-

sivity) fractures cannot be determined in the presence of high-flow zones. Consequently, if discrete-interval samples are required, if the transmissivity is lower than the measurement range of the flowmeter, or if all of the transmissive fractures in the borehole must be characterized for head and transmissivity, then BAT[3] testing is preferred.

Eleven boreholes were completed with semi-permanent DZM systems, which isolated two to five zones in the boreholes. Hydrographs constructed for discretely isolated zones in the boreholes showed seasonal changes in the magnitude of water levels and driving potential in response to precipitation and drought. Heads in discrete zones and in different boreholes vary in magnitude and timing in response to precipitation. In general, there are downward driving potentials in the recharge areas and in the area of the ground-water divide along the axis of the valley, and upward driving potentials in discharge areas north and south of the landfill.

Six boreholes were completed with continuous water-level monitoring systems in the DZM system. Continuous water-level records collected in selected boreholes show fluctuations that are not observed in the manual water-level records. The continuous records were analyzed to look for effects of pumping from a community well about 3,000 ft west of the landfill. No fluctuations that might indicate pumping or regular pumping cycles were identified in the continuous records. All of the continuous-record water-level data collected in open boreholes and in DZM systems showed a semi-diurnal pattern that coincides with gravimetric tidal plots generated for this area.

Cross-hole testing was done in the area of the former chemical-waste disposal pits. These tests entailed pumping from a single zone while monitoring the water levels in discrete zones in nearby boreholes completed in bedrock, in open-hole shallow bedrock wells, and in overburden wells. The lack of hydraulic response indicates a poor connection between the boreholes that were monitored.

Using hydrograph data, driving potential profiles were constructed to evaluate the north-south and east-west conceptual flowpaths in the study area. To further evaluate these flowpaths, the distribution of selected chemical constituents was analyzed relative to the interpreted driving potentials. The distribution of VOCs was consistent with the hydraulic distribution. Hydraulic heads from discrete zones were analyzed along a profile aligned with the axis of the north-south valley and extending from the minor ground-water divide near the chemical-waste disposal pits southward towards Eagleville Brook. The distribution of vertical heads along the profile indicate downward driving potentials near the ground-water divide, most often downward driving potentials but seasonally upward driving potentials at MW105R, and upward driving potentials at MW201R. A comparison of heads between the boreholes indicate that water-level altitudes are highest in the area of the ground-water divide and decline to the south. North of the ground-water divide, vertical driving potentials are downward at MW204R, upward in the top of MW103R, and upward at MW101R. A comparison of the heads between the boreholes indicate water-level altitudes that are highest in the area of the ground-water divide and decrease progressively to the north. These data indicate inferred flowpaths along the axis of the valley that are downward in the area of the ground-water divide and upward towards the surface-water discharge areas – the wetland to the north and the tributary to Eagleville Brook.

In the east-west direction, flow is driven by driving potentials from the hill east of the landfill. Localized shallow ground-water flow occurs along the flanks of the hill with downward driving potentials from recharge on the hill and upward driving potentials discharging toward the base of the hill slope. Additionally, some of the recharge on the hill is driven deeper into the aquifer and flows westward.

In the area of the ground-water divide near the center of the north-south valley and Hunting Lodge Road, hydraulic head and driving potential varies seasonally. During periods of recharge, there are downward driving potentials, and higher heads on the west side of the valley, which indicate potential flow towards the east. In the absence of recharge, there is less downward driving potential, and the heads in the area of the ground-water divide are as high as or higher than the heads in the boreholes to the west, indicating potential for westward flow. Even during periods of recharge when the highest driving potentials are in the east-west direction, the dominant driving potential is southward rather than westward.

Collectively, these results indicate that regional ground-water flow follows the topography. The local flow in bedrock, however, follows fractures that may be oriented differently than the regional flow directions. Additionally, these results show that local topographic highs exert control on the local flow in the bedrock aquifer in the area of the landfill.

The results of the study illustrate the importance of discrete-zone isolation and monitoring in fractured-rock aquifers to prevent cross contamination, to accurately measure hydraulic head at fracture zones of different depths, and to ensure collection of representative water-quality samples. The distribution of hydraulic heads and chemical concentrations were plotted to show the combined use of discrete-interval water-level and water-quality data for evaluating flow within the study area. The DZM systems provided the necessary discretization of heads and chemical concentrations to identify potential flow-paths and contaminant distribution. Moreover, the use of DZM systems prevented vertical flow through the open boreholes that could have spread the contamination and confused the interpretation of the distribution and source of contaminants at the site.

References Cited

Elci, Alper; Molz, F.J., III; and Waldrop, W.R., 2001, Implications of observed and simulated ambient flow in monitoring wells: Ground Water, v. 39, no. 6, p. 853-862.

Fahey, R.J. and Pease, M.H., Jr., 1977, Preliminary bedrock geologic map of the South Coventry quadrangle, Tolland county, Connecticut: U.S. Geological Survey Open-File Report 77-587, 30 p., 3 plates, scale 1:24,000.

Haley and Aldrich, Inc., Environmental Research Institute, Epona Associates, LLC, Regina Villa Associates, Inc., and Earth Tech, Inc., 2000, Final preliminary hydrogeologic investigation draft report, UConn landfill, Storrs, Conn.: Glastonbury, Connecticut, 267 p. plus attachments and appendices.

Haley and Aldrich, Inc.; Environmental Research Institute; Epona Associates, LLC; F.P. Haeni, LLC; and Regina Villa Associates, Inc.; with technical oversight by Mitretek Systems, Inc., 2002, Draft comprehensive hydrogeologic investigation and remedial action plan, University of Connecticut, Storrs, Connecticut: Glastonbury, Connecticut, File No. 91221-513.

Harrison, D.H., 1971, New computer programs for the calculation of earth tides: Cooperative Institute for Research in Environmental Sciences, NOAA/Univ. of Colorado, 29 p. (NNA.901211.0227)

Hsieh, P.A., Bredehoeft, J.D., and Farr, J.M., 1987, Determination of aquifer transmissivity from Earth tide analysis: Water Resources Research, v. 23, no. 10, p. 1824-1832.

Johnson, C.D., Haeni, F.P., Lane, J.W., and White, E.A., 2002, Borehole-geophysical investigation of the University of Connecticut landfill, Storrs, Connecticut: U.S. Geological Survey Water-Resources Investigations Report 01-4033, 187 p.

Johnson, C.D., and Kastrinos, J.R., 2002, Use of geophysical, hydraulic, and geochemical methods to develop a site conceptual ground-water flow model in central Connecticut, in Fractured Rock 2002, Denver, Colorado, March 13-15, 2002, Proceedings [abs.]: Westerville, Ohio, National Ground Water Association.

Johnson, C.D., Joesten, P.K., and Mondazzi, R.A., 2005, Borehole-geophysical and hydraulic investigation of the fractured-rock aquifer near the University of Connecticut landfill, Storrs, Connecticut, 2000 to 2001: U.S. Geological Survey Water-Resources Investigations Report 03-4125, 133 p.

National Oceanic and Atmospheric Administration, 2002, National Weather Service Forecast Office Boston, Massachusetts: National Oceanic and Atmospheric Administration web site at http://www.erh.noaa.gov/er/box/dailystns.html. (Accessed October 23, 2002.)

Paillet, F.L., 1998, Flow modeling and permeability estimation using borehole flow logs in heterogeneous fractured formations: Water Resources Research, v. 34, no. 5, p. 997-1010.

Paillet, F.L., 2000, A field technique for estimating aquifer parameters using flow log data: Ground Water, v. 38, no. 4, p. 510-521.

Parker, B.L., and Cherry, J.A., 1999, Scale considerations of chlorinated solvent source zones and contaminant fluxes—insights from detailed field studies, *in* Morganwalp, D.W., and Buxton, H.T., eds., U.S. Geological Survey Toxic Substance Hydrology Program - Proceedings of the Technical Meeting, Charleston, S.C., March 8-12, 1999: U.S. Geological Survey Water-Resources Investigations Report 99-4018C, v. 3, p. 3-5.

Powers, C.J., Wilson, Joanna, Haeni, F.P., and Johnson, C.D., 1999, Surface-geophysical investigation of the University of Connecticut landfill, Storrs, Connecticut: U.S. Geological Survey Water-Resources Investigations Report 99-4211, 34 p.

Shapiro, A.M., 2001, Characterizing ground-water chemistry and hydraulic properties of fractured rock aquifers using the multifunction bedrock-aquifer transportable testing tool (BAT^3): U.S. Geological Survey Fact Sheet; FS-075-01, 4 p.

Shapiro, A.M., 2002, Cautions and suggestions for geochemical sampling in fractured rock: Ground Water Monitoring and Remediation, v. 22, no. 3, p. 151-164.

Sterling, S.N., 1999, Comparison of discrete depth sampling using rock core and a removable multilevel system in a TCE contaminated fractured sandstone: MS Thesis, Department of Earth Sciences, University of Waterloo, Waterloo, Ontario, Canada, 108 p.

Todd, D.K., 1980, Groundwater hydrology (2d ed.): New York, N.Y., John Wiley and Sons, 535 p.

Williams, J.H., and Conger, R.W., 1990, Preliminary delineation of contaminated water-bearing fractures intersected by open-hole bedrock wells: Ground Water Monitoring Review, v. 10, no. 4, p. 118-126.

Winter, T. C., Harvey, J.W., Franke, O.L., and Alley, W. M., 1998, Ground water and surface water, a single resource: U.S. Geological Survey Circular 1139, 79 p.

Appendix 1

Calibration Values and Set Depths for Pressure Transducers in Boreholes in the UConn Landfill Study Area, Storrs, Connecticut

Appendix 1A. Calibration values and set depths for pressure transducers in MW101R.

[Z, zone in the discrete-zone monitoring system; C, channel in the continuous multi-channel tubing (CMT) where the transducer was placed; R^2, coefficient of determination showing the closeness of fit of the observed data and the line defined by the slope and offset; Shift, the distance that the transducer was moved from its original set depth; MP, measuring point, which is the top of the CMT]

Zone	Slope	Offset	R^2	Shift	Dates applicable	Transducer	Elevation of MP, in feet	Set depth, in feet
Z1, C1	3.296	-2.756	0.986	2.6	1/1/02 to 6/18/02	1	553.84	14.6
Z1, C1	3.296	-2.756	0.986	2.45	6/18/02 to 8/20/02	1	553.84	14.6
Z1, C1	3.296	-2.756	0.986	2.8	8/21/02 to 9/26/02	1	553.84	14.6
Z1, C1	3.296	-2.756	0.986	2.5	9/26/02 to 1/1/03	1	553.84	14.6
Z2, C3	2.399	-0.160	1.000	0	1/1/02 to 9/26/02	2	553.84	15.0
Z2, C3	2.193	0.744	0.997	0	9/26/02 to 1/1/03	2	553.84	15.0
Z3, C5	2.129	0.725	0.998	0	1/1/02 to 6/18/02	3	553.84	15.0
Z3, C5	2.129	0.725	0.998	2.6	9/23/02 to 9/26/02	3	553.84	15.0
Z3, C5	2.264	2.914	1.000	0.2	9/26/02 to 10/15/02	3	553.84	15.0
Z3, C5	2.266	0.262	1.000	0	10/15/02 to 1/1/03	3	553.84	9.8

Appendix 1B. Calibration values and set depths for pressure transducers in MW103R.

[Z, zone in the discrete-zone monitoring system; C, channel in the continuous multi-channel tubing (CMT) where the transducer was placed; C1/2, 0.5-inch diameter tube suspended in upper zone where the transducer was placed; R^2, coefficient of determination showing the closeness of fit of the observed data and the line defined by the slope and offset; Shift, the distance that the transducer was moved from its original set depth; MP, measuring point, which is the top of the CMT]

Zone	Slope	Offset	R^2	Shift	Dates applicable	Transducer	Elevation of MP, in feet	Set depth, in feet
Z1, C1/2	2.2814	0.2126	0.9999	0	1/1/02 to 1/1/03	1	571.37	20
Z2, C2	2.3374	0.1783	1.0000	0	1/1/02 to 1/1/03	2	571.40	19.8
Z3, C4	2.3253	0.5215	0.9761	0	1/1/02 to 1/1/03	3	571.40	20
Z5, C6	2.3441	-3.8767	0.9999	0	1/1/02 to 1/1/03	4	571.40	20

Appendix 1C. Calibration values and set depths for pressure transducers in MW104R.

[Z, zone in the discrete-zone monitoring system; C, channel in the continuous multi-channel tubing (CMT) where the transducer was placed; R^2, coefficient of determination showing the closeness of fit of the observed data and the line defined by the slope and offset; Shift, the distance that the transducer was moved from its original set depth; MP, measuring point, which is the top of the CMT]

Zone	Slope	Offset	R^2	Shift	Dates applicable	Transducer	Elevation of MP, in feet	Set depth, in feet
Z1-C1	2.405	0.370	0.9996	-0.1	1/1/02 to 2/1/02	1	575.71	30.5
Z1-C1	2.405	0.370	0.9996	-0.4	2/1/02 to 7/12/02	1	575.71	30.5
Z1-C1	2.405	0.370	0.9996	-0.1	7/12/02 to 1/1/03	1	575.71	30.5
Z3-C4	2.451	-0.458	0.9997	0	1/1/02 to 1/1/03	2	575.71	35
Z5-C7	2.575	-1.574	0.9974	0	1/1/02 to 2/1/02	3	575.71	38
Z5-C7	2.575	-1.574	0.9974	-0.3	2/1/02 to 4/26/02	3	575.71	38
Z5-C7	2.575	-1.574	0.9974	-0.5	4/26/02 to 8/22/02	3	575.71	38
Z5-C7	2.575	-1.574	0.9974	-0.4	8/22/02 to 1/1/03	3	575.71	38

Appendix 1D. Calibration values and set depths for pressure transducers in MW105R.

[Z, zone in the discrete-zone monitoring system; C, channel in the continuous multi-channel tubing (CMT) where the transducer was placed; C1/2, 0.5-inch diameter tube suspended in upper zone where the transducer was placed; R^2, coefficient of determination showing the closeness of fit of the observed data and the line defined by the slope and offset; Shift, the distance that the transducer was moved from its original set depth; MP, measuring point, which is the top of the CMT]

Zone	Slope	Offset	R^2	Shift	Dates applicable	Transducer	Elevation of MP, in feet	Set depth, in feet
Z1, C1/2	2.411	-1.179	0.9994	0	1/1/02 to 2/1/02	1	565.82	30
Z1, C1/2	2.411	-1.179	0.9994	-0.3	2/1/02 to 1/1/03	1	565.82	30
Z2, C1	2.404	0.335	0.9937	0	1/1/02 to 11/26/02	2	565.91	25
Z2, C1	2.404	0.335	0.9937	-0.2	11/26/02 to 1/1/03	2	565.91	25
Z3, C4	2.326	-0.071	0.9936	0	1/1/02 to 1/1/03	3	565.91	25
Z5, C7	2.470	0.197	0.9719	0	1/1/02 to 4/16/02	4	565.91	19
Z5, C7	2.470	0.197	0.9719	-0.3	4/16/02 to 7/15/02	4	565.91	19
Z5, C7	2.470	0.197	0.9719	0	7/15/02 to 1/1/03	4	565.91	19

Appendix 1E. Calibration values and set depths for pressure transducers in MW109R.

[Z, zone in the discrete-zone monitoring system; C, channel in the continuous multi-channel tubing (CMT) where the transducer was placed; C0, 6-inch diameter casing where the transducer was placed; R^2, coefficient of determination showing the closeness of fit of the observed data and the line defined by the slope and offset; Shift, the distance that the transducer was moved from its original set depth; MP, measuring point, which is the top of the CMT]

Zone	Slope	Offset	R^2	Shift	Dates applicable	Transducer	Elevation of MP, in feet	Set depth, in feet
Z1, C0	2.353	0.526	0.9799	0	1/1/02 to 8/11/02	1	582.69	30.4
Z1, C0	2.353	0.526	0.9799	0.6	8/11/02 to 1/1/03	1	582.69	30.4
Z2, C2	2.747	-0.485	0.9757	0	1/1/02 to 3/9/02	2	582.77	30.1
Z2, C2	2.339	0.566	0.9998	0	3/9/02 to 1/1/03	2	582.77	30.1
Z3, C4	2.403	-0.691	0.9579	0	1/1/02 to 1/14/02	3	582.77	30.1
Z3, C4	2.403	-0.691	0.9579	0.1	1/14/02 to 3/20/02	3	582.77	30.1
Z3, C4	2.403	-0.691	0.9579	0.2	3/20/02 to 6/18/02	3	582.77	30.1
Z3, C4	2.403	-0.691	0.9579	0	6/18/02 to 1/1/03	3	582.77	30.1
Z4, C6	2.372	-0.433	0.9792	0	1/1/02 to 1/1/03	4	582.77	30.1

Appendix 1F. Calibration values and set depths for pressure transducers in MW122R.

[Z, zone in the discrete-zone monitoring system; C, channel in the continuous multi-channel tubing (CMT) where the transducer was placed; R^2, coefficient of determination showing the closeness of fit of the observed data and the line defined by the slope and offset; Shift, the distance that the transducer was moved from its original set depth; MP, measuring point, which is the top of the CMT]

Zone	Slope	Offset	R^2	Shift	Dates applicable	Transducer	Elevation of MP, in feet	Set depth, in feet
Z1	2.3092	-0.0922	1.0000	0	1/1/02 to 1/1/03	1	581.76	30
Z2	2.2912	0.0476	0.9999	0	1/1/02 to 1/1/03	2	581.76	32

Appendix 2

Open-Hole and Discrete-Zone Manual Water-Level Measurements in Boreholes in the UConn Landfill Study Area, Storrs, Connecticut

Appendix 2A. Open-hole and discrete-zone manual water-level measurements in MW101R.

[Depth is depth to water, in feet. Elevation (elev.) is elevation of head, in feet above NAVD 88. "Z" means Zone in borehole. * means measurements referenced to 6-inch steel casing. All other measurements referenced to continuous multi-channel tubing. Numbers with parentheses indicates elevation of open zone in feet above North American Vertical Datum of 1988 (NAVD 88). Packer pressure in pounds force per square inch. ft, feet. NA, not applicable. NR, not recorded. I, inflated. L, low pressure. FZ, frozen. --, missing head data]

Date	Time	MW101R*		Z1* (530.5-514.5 ft)		Z1 (530.5-514.5 ft)		Z2 (512.4-493.5 ft)		Z3 (491.5-485.0 ft)		Z4 (483.0-426.5 ft)		Packer pressure	Packer condition
		Depth	Elev.	Depth	Elev.	Depth	Elev.	Depth	Elev.	Depth	Elev.	Depth	Elev.		
5/31/00	12:00	2.60	550.93	--	--	--	--	--	--	2.83	551.09	2.86	551.06	NR	NA
6/9/00	17:00	--	--	2.40	551.13	2.72	551.20	2.82	551.10	1.84	552.08	1.83	552.09	10	I
6/12/00	10:00	--	--	2.75	550.78	3.21	550.71	2.74	551.18	1.84	552.08	0.66	553.26	0	L
6/19/00	12:38	--	--	3.12	550.41	3.54	550.38	2.90	551.02	1.90	552.02	1.72	552.20	13	I
6/20/00	12:00	--	--	3.26	550.27	3.62	550.30	2.98	550.94	--	--	--	--	12	I
6/26/00	12:00	--	--	--	--	3.68	550.24	3.56	550.36	--	--	--	--	NR	NR
6/27/00	8:25	--	--	3.57	549.96	3.94	549.98	3.22	550.70	2.13	551.79	2.03	551.89	14	I
7/7/00	9:20	--	--	3.60	549.93	3.96	549.96	3.29	550.63	2.20	551.72	1.98	551.94	10	I
6/1/00	10:30	--	--	3.56	549.97	3.93	549.99	3.31	550.61	2.22	551.70	2.05	551.87	15	I
7/31/00	12:00	--	--	--	--	3.00	550.92	2.85	551.07	2.77	551.15	1.98	551.94	NR	NR
8/1/00	13:15	--	--	3.06	550.47	3.42	550.50	2.85	551.07	1.76	552.16	1.59	552.33	13	I
8/8/00	10:30	--	--	3.61	549.92	3.96	549.96	3.43	550.49	2.26	551.66	1.54	552.38	13	I
8/17/00	12:00	--	--	--	--	3.53	550.39	3.44	550.48	2.26	551.66	2.06	551.86	NR	NR
8/28/00	10:35	--	--	3.75	549.78	4.16	549.76	3.97	549.95	2.80	551.12	2.09	551.83	8	I
9/20/00	11:30	--	--	2.94	550.59	3.33	550.59	3.15	550.77	2.22	551.70	2.67	551.25	8	NR
9/28/00	12:00	--	--	--	--	3.79	550.13	3.68	550.24	2.80	551.12	2.13	551.79	NR	I
9/29/00	10:00	--	--	3.58	549.95	3.98	549.94	3.77	550.15	2.80	551.12	2.65	551.27	14	I
10/6/00	14:45	--	--	3.62	549.91	4.00	549.92	3.81	550.11	2.82	551.10	2.62	551.30	9	I
10/20/00	9:00	--	--	3.73	549.80	4.11	549.81	3.92	550.00	3.09	550.83	2.80	551.12	6	I
11/2/00	10:07	--	--	3.93	549.60	4.31	549.61	4.13	549.79	3.31	550.61	3.02	550.90	4	L
11/27/00	9:53	--	--	3.29	550.24	3.71	550.21	3.53	550.39	2.64	551.28	3.27	550.65	5	L
11/29/00	8:40	3.47	550.06	3.59	549.94	3.93	549.91	3.68	550.16	2.94	550.90	2.53	551.39	NR	NA
11/30/00	9:39	--	--	--	--	--	--	--	--	2.89	550.95	2.71	551.13	16	I
12/1/00	12:00	--	--	--	--	4.03	549.71	--	--	3.04	550.80	2.69	551.15	NR	NR
12/3/00	9:00	--	--	3.68	549.85	4.15	549.69	3.78	550.06	3.20	550.64	2.77	551.07	NR	NR
12/11/00	10:30	--	--	3.81	549.72	3.71	550.13	3.97	549.87	3.67	550.17	2.98	550.86	9	I
12/14/00	12:00	--	--	3.35	550.18	3.76	550.08	3.72	550.12	3.71	550.13	3.66	550.18	NR	FZ
12/15/00	9:20	--	--	3.41	550.12	3.16	550.68	3.74	550.10	2.28	551.56	3.72	550.12	NR	FZ
12/18/00	9:25	--	--	2.83	550.70	3.44	550.40	3.07	550.77	2.52	551.32	1.98	551.86	15	I
12/21/00	10:45	--	--	3.11	550.42	3.77	550.07	3.32	550.52	2.71	551.13	2.24	551.60	NR	FZ
1/3/01	9:50	--	--	3.43	550.10	3.86	549.98	3.63	550.21	2.68	551.16	2.45	551.39	NR	FZ
1/30/01	12:00	--	--	--	--	3.27	550.57	3.58	550.26	2.24	551.60	2.54	551.30	NR	FZ
2/1/01	16:12	--	--	2.90	550.63	3.27	550.57	3.15	550.69	2.16	551.68	2.02	551.82	NR	FZ
2/13/01	10:25	--	--	2.91	550.62	2.91	550.93	3.14	550.70	1.76	552.08	1.90	551.94	NR	FZ
2/28/01	12:00	--	--	--	--	2.79	551.05	2.81	551.03	1.58	552.26	1.51	552.33	NR	FZ
3/14/01	11:10	--	--	2.49	551.04	--	--	2.64	551.20	1.48	552.36	1.27	552.57	NR	FZ
4/5/01	11:12	--	--	2.38	551.15	--	--	2.53	551.31	--	--	1.18	552.66	NR	NR

Appendix 2A. Open-hole and discrete-zone manual water-level measurements in MW101R.—Continued

[Depth is depth to water, in feet. Elevation (elev.) is elevation of head, in feet above NAVD 88. "Z" means Zone in borehole. * means measurements referenced to 6-inch steel casing. All other measurements referenced to continuous multi-channel tubing. Numbers with parentheses indicates elevation of open zone in feet above North American Vertical Datum of 1988 (NAVD 88). Packer pressure in pounds force per square inch. ft, feet. NA, not applicable. NR, not recorded. I, inflated. L, low pressure. FZ, frozen. --, missing head data]

Date	Time	MW101R* Depth	MW101R* Elev.	Z1* (530.5-514.5 ft) Depth	Z1* Elev.	Z1 (530.5-514.5 ft) Depth	Z1 Elev.	Z2 (512.4-493.5 ft) Depth	Z2 Elev.	Z3 (491.5-485.0 ft) Depth	Z3 Elev.	Z4 (483.0-426.5 ft) Depth	Z4 Elev.	Packer pressure	Packer condition
4/12/01	12:00	--	--	--	--	--	--	2.35	551.49	1.26	552.58	0.89	552.95	NR	NR
4/20/01	9:44	--	--	2.61	550.92	--	--	2.72	551.12	1.66	552.18	1.32	552.52	NR	NR
5/1/01	10:58	--	--	2.66	550.87	--	--	2.78	551.06	1.67	552.17	1.25	552.59	NR	NR
5/16/01	10:50	--	--	3.04	550.49	--	--	3.16	550.68	2.04	551.80	1.62	552.22	NR	NR
6/4/01	9:49	--	--	2.50	551.03	--	--	2.65	551.19	1.48	552.36	1.03	552.81	NR	NR
6/4/01	12:00	--	--	--	--	--	--	2.65	551.19	1.48	552.36	1.03	552.81	NR	NR
6/25/01	9:53	--	--	2.78	550.75	--	--	2.94	550.90	1.75	552.09	1.29	552.55	NR	NR
7/10/01	9:30	--	--	3.24	550.29	--	--	3.35	550.49	2.16	551.68	1.68	552.16	NR	NR
7/24/01	9:50	--	--	3.66	549.87	--	--	3.78	550.06	2.57	551.27	2.10	551.74	NR	NR
8/13/01	9:14	--	--	3.32	550.21	--	--	3.51	550.33	2.48	551.36	2.21	551.63	NR	NR
9/5/01	1:15	--	--	4.40	549.13	--	--	4.23	549.61	3.41	550.43	--	--	NR	I
9/11/01	3:18	--	--	4.28	549.25	--	--	4.43	549.41	3.51	550.33	3.08	550.76	12	I
9/27/01	0:20	--	--	3.65	549.88	--	--	3.86	549.98	3.05	550.79	2.67	551.17	11	I
10/9/01	0:15	--	--	4.13	549.40	--	--	4.31	549.53	3.48	550.36	3.13	550.71	10	I
10/12/01	5:16	--	--	4.07	549.46	--	--	4.28	549.56	3.41	550.43	3.05	550.79	17	I
10/30/01	9:08	--	--	4.31	549.22	--	--	4.49	549.35	3.68	550.16	3.27	550.57	14	I
11/15/01	1:09	--	--	4.32	549.21	--	--	4.51	549.33	3.75	550.09	3.41	550.43	7	I
11/29/01	11:02	--	--	4.37	549.16	--	--	4.56	549.28	3.83	550.01	3.52	550.32	15	I
12/21/01	12:06	--	--	3.92	549.61	--	--	4.15	549.69	3.43	550.41	3.08	550.76	12	I
1/14/02	11:19	--	--	3.80	549.73	--	--	4.08	549.76	3.93	549.91	3.94	549.90	NR	FZ
1/31/02	11:16	--	--	3.74	549.79	--	--	3.98	549.86	3.75	550.09	3.77	550.07	NR	FZ
2/28/02	10:21	--	--	3.72	549.81	--	--	3.99	549.85	3.78	550.06	3.78	550.06	NR	FZ
3/26/02	11:12	--	--	3.20	550.33	--	--	3.48	550.36	3.28	550.56	3.27	550.57	NR	FZ
4/16/02	11:30	--	--	3.02	550.51	--	--	3.29	550.55	3.02	550.82	3.02	550.82	0	L
5/7/02	13:18	--	--	2.98	550.55	--	--	3.15	550.69	2.33	551.51	1.93	551.91	17	I
6/18/02	12:55	--	--	2.84	550.69	--	--	3.00	550.84	1.92	551.92	1.45	552.39	17	I
8/21/02	12:10	--	--	4.22	549.31	--	--	4.43	549.41	3.23	550.61	2.86	550.98	14	I
9/19/02	11:15	--	--	4.26	549.27	--	--	4.45	549.39	3.44	550.40	3.28	550.56	20	I
9/26/02	9:21	--	--	4.37	549.16	--	--	4.57	549.27	3.77	550.07	3.39	550.45	18	I
10/10/02	8:40	--	--	4.38	549.15	--	--	4.57	549.27	3.75	550.09	3.39	550.45	18	I
10/15/02	10:18	--	--	3.97	549.56	--	--	4.17	549.67	3.37	550.47	3.04	550.80	17	I
10/28/02	10:29	--	--	3.55	549.98	--	--	3.77	550.07	2.87	550.97	2.54	551.30	15	I
11/26/02	11:48	--	--	3.20	550.33	--	--	3.38	550.46	2.59	551.25	2.21	551.63	11	I
12/3/02	15:35	--	--	3.38	550.15	--	--	3.57	550.27	2.77	551.07	2.32	551.52	NR	FZ

Appendix 2B. Open-hole and discrete-zone manual water-level measurements in MW103R.

[Depth is depth to water, in feet. Elevation (elev.) is elevation of head, in feet above NAVD 88. "Z" means Zone in borehole. * means measurements referenced to 6-inch steel casing. All other measurements referenced to continuous multi-channel tubing. Numbers with parentheses indicates elevation of open zone in feet above North American Vertical Datum of 1988 (NAVD 88). Packer pressure in pounds force per square inch. ft, feet. NA, not applicable. NR, not recorded. I, inflated. L, low pressure. FZ, frozen. --, missing head data]

Date	Time	MW103R* Depth	MW103R* Elev.	Z1* (544.5-524.0 ft) Depth	Z1* (544.5-524.0 ft) Elev.	Z1 (544.5-524.0 ft) Depth	Z1 (544.5-524.0 ft) Elev.	Z2 (522.0-495.0 ft) Depth	Z2 (522.0-495.0 ft) Elev.	Z3 (493.0-483.0 ft) Depth	Z3 (493.0-483.0 ft) Elev.	Z4 (481.0-460.0 ft) Depth	Z4 (481.0-460.0 ft) Elev.	Z5 (458.0-440.0 ft) Depth	Z5 (458.0-440.0 ft) Elev.	Packer pressure	Packer condition
9/12/00	13:00	8.62	562.38	--	--	--	--	--	--	--	--	--	--	--	--	NR	NA
9/13/00	9:36	--	--	--	--	--	--	8.33	563.01	8.19	563.15	8.28	563.06	8.38	562.96	11	I
9/14/00	13:55	--	--	8.86	562.14	--	--	8.39	562.95	8.35	562.99	8.26	563.08	8.43	562.91	8	I
9/20/00	8:20	--	--	6.50	564.50	7.29	563.99	7.57	563.77	7.54	563.80	7.51	563.83	7.62	563.72	6	I
9/21/00	13:51	--	--	--	--	--	--	6.99	564.35	6.84	564.50	6.88	564.46	7.03	564.31	11	I
9/22/00	15:40	--	--	7.87	563.13	8.21	563.07	7.07	564.27	30.57	540.77	7.08	564.26	7.15	564.19	0	L
9/25/00	10:25	--	--	7.72	563.28	8.05	563.23	7.18	564.16	6.97	564.37	7.03	564.31	7.23	564.11	14	I
9/28/00	12:00	--	--	--	--	--	--	7.86	563.48	7.85	563.49	7.85	563.49	7.84	563.50	NR	NR
9/29/00	10:20	--	--	7.61	563.39	7.93	563.35	7.96	563.38	7.96	563.38	7.96	563.38	7.97	563.37	2	L
10/19/00	7:55	8.80	562.20	--	--	--	--	--	--	--	--	--	--	--	--	NR	NA
10/19/00	10:03	--	--	7.74	563.26	9.08	562.29	8.77	562.63	8.19	563.21	7.50	563.90	8.81	562.59	20	I
10/20/00	9:30	--	--	9.15	561.85	9.54	561.83	8.66	562.74	8.68	562.72	8.52	562.88	8.69	562.71	14	I
10/27/00	11:35	--	--	9.36	561.64	9.73	561.64	8.79	562.61	8.68	562.72	8.62	562.78	8.83	562.57	10	I
11/2/00	10:52	--	--	9.58	561.42	9.93	561.44	8.97	562.43	8.89	562.51	8.80	562.60	8.99	562.41	14	I
11/27/00	10:29	--	--	7.10	563.90	--	--	7.46	563.94	7.39	564.01	7.29	564.11	7.49	563.91	10	I
12/1/00	12:00	--	--	--	--	--	--	7.13	564.27	7.12	564.28	7.10	564.30	--	--	NR	NR
12/11/00	10:54	--	--	7.68	563.32	8.07	563.30	7.12	564.28	7.08	564.32	6.86	564.54	7.16	564.24	15	I
12/18/00	15:35	--	--	6.31	564.69	6.71	564.66	8.81	562.59	8.42	562.98	8.30	563.10	8.68	562.72	14	I
12/21/00	11:12	--	--	7.51	563.49	5.87	565.50	3.88	567.52	3.92	567.48	3.84	567.56	3.90	567.50	NR	FZ
1/2/01	12:37	--	--	5.78	565.22	6.12	565.25	4.35	567.05	4.32	567.08	4.16	567.24	4.40	567.00	NR	FZ
1/3/01	10:10	--	--	5.80	565.2	6.13	565.24	4.37	567.03	4.38	567.02	4.33	567.07	4.40	567.00	NR	FZ
1/5/01	9:25	--	--	5.76	565.24	6.08	565.29	4.42	566.98	4.41	566.99	4.36	567.04	4.46	566.94	NR	FZ
1/8/01	9:50	--	--	5.87	565.13	6.14	565.23	4.61	566.79	4.59	566.81	4.50	566.90	4.61	566.79	NR	FZ
1/30/01	12:00	--	--	--	--	5.99	565.38	5.15	566.25	5.14	566.26	4.99	566.41	5.20	566.20	NR	FZ
2/1/01	16:25	--	--	5.65	565.35	5.91	565.46	4.52	566.88	4.54	566.86	4.49	566.91	4.57	566.83	NR	FZ
2/13/01	10:45	--	--	5.56	565.44	--	--	4.12	567.28	4.14	567.26	4.04	567.36	4.16	567.24	NR	FZ
2/28/01	12:00	--	--	--	--	--	--	3.76	567.64	3.75	567.65	3.59	567.81	3.61	567.79	NR	FZ
3/14/01	10:45	--	--	5.18	565.82	5.52	565.85	3.57	567.83	3.53	567.87	3.28	568.12	3.63	567.77	NR	FZ
3/19/01	10:05	--	--	--	--	5.35	566.02	--	--	2.87	568.53	2.57	568.83	--	--	NR	FZ
4/5/01	11:41	--	--	4.89	566.11	--	--	2.09	569.31	2.09	569.31	--	--	2.14	569.26	14	I
4/20/01	10:00	--	--	5.15	565.85	--	--	3.03	568.37	2.97	568.43	--	--	3.04	568.36	NR	NR
4/26/01	12:00	--	--	--	--	--	--	3.08	568.32	3.08	568.32	--	--	--	--	NR	NR
5/1/01	11:17	--	--	5.33	565.67	--	--	3.52	567.88	3.56	567.84	--	--	3.56	567.84	NR	NR

Appendix 2B. Open-hole and discrete-zone manual water-level measurements in MW103R.—Continued

[Depth is depth to water, in feet. Elevation (elev.) is elevation of head, in feet above NAVD 88. "Z" means Zone in borehole. * means measurements referenced to 6-inch steel casing. All other measurements referenced to continuous multi-channel tubing. Numbers with parentheses indicates elevation of open zone in feet above North American Vertical Datum of 1988 (NAVD 88). Packer pressure in pounds force per square inch. ft, feet. NA, not applicable. NR, not recorded. I, inflated. L, low pressure. FZ, frozen. --, missing head data]

Date	Time	MW103R+ Depth	MW103R+ Elev.	Z1* (544.5-524.0 ft) Depth	Z1* (544.5-524.0 ft) Elev.	Z1 (544.5-524.0 ft) Depth	Z1 (544.5-524.0 ft) Elev.	Z2 (522.0-495.0 ft) Depth	Z2 (522.0-495.0 ft) Elev.	Z3 (493.0-483.0 ft) Depth	Z3 (493.0-483.0 ft) Elev.	Z4 (481.0-460.0 ft) Depth	Z4 (481.0-460.0 ft) Elev.	Z5 (458.0-440.0 ft) Depth	Z5 (458.0-440.0 ft) Elev.	Packer pressure	Packer condition
5/16/01	11:12	--	--	6.02	564.98	--	--	4.77	566.63	4.77	566.63	--	--	4.77	566.63	NR	NR
6/4/01	10:12	--	--	5.44	565.56	--	--	3.92	567.48	3.36	567.54	--	--	3.92	567.48	NR	NR
6/4/01	12:00	--	--	--	--	--	--	3.94	567.46	3.89	567.51	--	--	3.97	567.43	NR	NR
6/25/01	10:08	--	--	5.50	565.5	--	--	3.59	567.81	3.48	567.92	--	--	3.62	567.78	NR	NR
7/10/01	9:50	--	--	6.76	564.24	--	--	5.47	565.93	5.31	566.09	--	--	5.49	565.91	NR	NR
7/24/01	11:25	--	--	8.46	562.54	9.43	561.94	7.22	564.18	7.03	564.37	--	--	7.27	564.13	NR	NR
8/13/01	9:31	--	--	9.11	561.89	--	--	8.94	562.46	8.82	562.58	8.81	562.59	8.93	562.47	NR	NR
8/30/01	12:45	--	--	9.63	561.37	--	--	--	--	9.32	562.08	9.28	562.12	9.29	562.11	NR	NR
9/11/01	10:25	--	--	10.46	560.54	--	--	10.05	561.35	10.10	561.30	10.01	561.39	10.09	561.31	18	I
9/27/01	11:05	--	--	9.37	561.63	--	--	9.65	561.75	9.62	561.78	9.50	561.9	9.74	561.66	17	I
10/9/01	10:38	--	--	10.40	560.60	--	--	10.27	561.13	10.26	561.14	10.11	561.29	10.28	561.12	17	I
10/30/01	9:21	--	--	11.44	559.56	--	--	11.24	560.16	11.11	560.29	10.95	560.45	11.24	560.16	15	I
11/15/01	11:32	--	--	12.00	559.00	--	--	11.67	559.73	11.49	559.91	11.40	560.00	11.66	559.74	15	I
11/29/01	11:15	--	--	12.47	558.53	--	--	11.94	559.46	11.83	559.57	11.74	559.66	12.02	559.38	7	I
12/27/01	11:50	--	--	10.04	560.96	--	--	10.16	561.24	10.00	561.40	9.96	561.44	10.22	561.18	NR	FZ
1/14/02	11:50	--	--	9.68	561.32	--	--	9.90	561.50	9.77	561.63	9.79	561.61	9.97	561.43	NR	FZ
1/31/02	11:39	--	--	8.80	562.20	--	--	9.05	562.35	8.85	562.55	8.87	562.53	9.10	562.30	NR	FZ
2/7/02	11:40	--	--	8.45	562.55	--	--	8.50	562.90	8.50	562.90	8.51	562.89	8.52	562.88	NR	FZ
2/7/02	13:00	--	--	8.42	562.58	--	--	8.48	562.92	7.88	563.52	7.55	563.85	8.51	562.89	21	I
2/28/02	11:15	--	--	8.04	562.96	--	--	7.89	563.51	7.75	563.65	7.68	563.72	7.94	563.46	13	I
3/28/02	9:52	--	--	5.78	565.22	--	--	4.90	566.50	4.66	566.74	4.55	566.85	4.95	566.45	17	I
4/18/02	10:20	--	--	5.77	565.23	--	--	4.38	567.02	4.15	567.25	4.04	567.36	4.44	566.96	8	I
5/7/02	11:10	--	--	5.58	565.42	--	--	3.99	567.41	--	--	3.91	567.49	4.03	567.37	16	I
6/18/02	13:22	--	--	5.63	565.37	--	--	3.89	567.51	3.72	567.68	3.67	567.73	3.92	567.48	17	I
7/10/02	9:05	--	--	8.21	562.79	--	--	6.81	564.59	6.61	564.79	6.54	564.86	6.84	564.56	16	I
8/21/02	11:53	--	--	10.79	560.21	--	--	10.09	561.31	9.82	561.58	9.74	561.66	10.13	561.27	13	I
9/24/02	13:55	--	--	11.28	559.72	--	--	10.89	560.51	10.91	560.49	10.91	560.49	10.94	560.46	18	I
10/28/02	11:23	--	--	9.13	561.87	--	--	9.44	561.96	9.44	561.96	9.35	562.05	9.47	561.93	17	I
11/26/02	12:30	--	--	6.65	564.35	--	--	6.29	565.11	6.20	565.20	6.09	565.31	6.33	565.07	17	I

Appendix 2C. Open-hole and discrete-zone manual water-level measurements in MW104R.

[Depth is depth to water, in feet. Elevation (elev.) is elevation of head, in feet above NAVD 88. "Z" means Zone in borehole. * means measurements referenced to 6-inch steel casing. All other measurements referenced to continuous multi-channel tubing. Numbers with parentheses indicates elevation of open zone in feet above North American Vertical Datum of 1988 (NAVD 88). Packer pressure in pounds force per square inch. ft, feet. NA, not applicable. NR, not recorded. I, inflated. L, low pressure. FZ, frozen. --, missing head data]

Date	Time	MW104R* Depth	MW104R* Elev.	Z1* (557.5-539.5 ft) Depth	Z1* Elev.	Z1 (557.5-539.5 ft) Depth	Z1 Elev.	Z2 (537.5-503.5 ft) Depth	Z2 Elev.	Z3 (501.5-491.5 ft) Depth	Z3 Elev.	Z4 (489.5-472.5 ft) Depth	Z4 Elev.	Z5 (470.5-448.5 ft) Depth	Z5 Elev.	Packer pressure	Packer condition
9/18/00	11:10	19.00	556.46	--	--	--	--	--	--	--	--	--	--	--	--	NR	NA
9/18/00	9:00	--	--	12.99	562.47	13.23	562.48	17.79	557.92	18.52	557.19	19.92	555.79	19.98	555.73	12	I
9/20/00	9:30	--	--	11.94	--	12.21	563.50	17.46	558.25	18.30	557.41	19.73	555.98	19.78	555.93	24	I
9/28/00	12:00	--	--	--	--	12.44	563.27	17.30	558.41	18.10	557.61	19.57	556.14	19.58	556.13	NR	NR
9/29/00	11:10	--	--	12.41	563.05	12.57	563.14	17.35	558.36	18.20	557.51	19.65	556.06	19.65	556.06	16	I
10/6/00	15:32	--	--	13.06	562.40	13.28	562.43	17.68	558.03	18.49	557.22	19.86	555.85	19.92	555.79	14	I
10/20/00	9:45	--	--	13.82	561.64	14.05	561.66	18.45	557.26	19.33	556.38	20.69	555.02	20.71	555.00	17	I
10/27/00	12:00	--	--	14.38	561.08	14.60	561.11	18.68	557.03	19.58	556.13	20.90	554.81	20.92	554.79	17	I
11/2/00	11:26	--	--	14.63	560.83	14.85	560.86	18.93	556.78	19.80	555.91	21.07	554.64	21.10	554.61	16	I
11/27/00	11:40	--	--	12.75	562.71	12.94	562.77	18.57	557.14	19.37	556.34	20.54	555.17	20.57	555.14	12	I
12/11/00	11:20	--	--	13.55	561.91	13.78	561.93	18.42	557.29	19.28	556.43	20.37	555.34	20.39	555.32	18	I
12/21/00	11:49	--	--	9.42	566.04	9.60	566.11	16.56	559.15	17.23	558.48	18.43	557.28	18.47	557.24	7	I
1/3/01	11:05	--	--	10.05	565.41	10.25	565.46	16.18	559.53	17.11	558.60	18.37	557.34	18.39	557.32	NR	FZ
1/30/01	12:00	--	--	--	--	--	--	17.37	558.34	18.01	557.70	19.27	556.44	19.29	556.42	NR	FZ
2/1/01	17:00	--	--	10.03	565.43	10.26	565.45	16.65	559.06	17.72	557.99	18.98	556.73	19.11	556.60	NR	FZ
2/13/01	11:29	--	--	9.26	566.20	9.49	566.22	15.15	560.56	16.95	558.76	18.14	557.57	18.28	557.43	NR	FZ
2/28/01	12:00	--	--	--	--	--	--	12.91	562.80	--	--	--	--	--	--	NR	FZ
3/14/01	10:05	--	--	7.88	567.58	8.07	567.64	9.82	565.89	15.75	559.96	17.17	558.54	17.29	558.42	NR	FZ
3/16/01	9:45	--	--	--	--	7.58	568.13	9.33	566.38	9.45	566.26	10.08	565.63	17.04	558.67	NR	FZ
4/5/01	12:58	--	--	6.08	569.38	--	--	--	--	9.06	566.65	--	--	13.83	561.88	NR	NR
4/20/01	8:30	--	--	7.25	568.21	--	--	--	--	10.52	565.19	--	--	14.95	560.76	NR	NR
4/26/01	12:00	--	--	7.25	568.21	--	--	--	--	10.52	565.19	--	--	14.95	560.76	NR	NR
5/1/01	11:57	--	--	8.18	567.28	--	--	--	--	11.65	564.06	--	--	15.67	560.04	NR	NR
5/16/01	12:06	--	--	9.66	565.80	--	--	--	--	13.28	562.43	--	--	16.95	558.76	NR	NR
6/4/01	11:00	--	--	8.37	567.09	--	--	--	--	12.46	563.25	--	--	16.39	559.32	NR	NR
6/25/01	10:58	--	--	8.28	567.18	--	--	--	--	12.43	563.28	--	--	15.96	559.75	NR	NR
7/10/01	10:39	--	--	10.31	565.15	--	--	--	--	14.79	560.92	--	--	17.72	557.99	NR	NR
7/24/01	12:15	--	--	12.36	563.10	--	--	--	--	16.96	558.75	--	--	19.22	556.49	NR	NR
8/13/01	10:35	--	--	14.08	561.38	14.32	561.39	14.73	560.98	18.69	557.02	20.41	555.30	20.86	554.85	NR	NR
8/30/01	12:00	--	--	14.57	560.89	--	--	15.21	560.50	19.26	556.45	20.87	554.84	21.31	554.40	NR	NR
9/11/01	12:30	--	--	15.45	560.01	--	--	16.12	559.59	19.97	555.74	21.52	554.19	21.95	553.76	0	L
9/18/01	15:00	--	--	--	--	--	--	--	--	--	--	--	--	--	--	25	I
9/27/01	12:00	--	--	13.77	561.69	--	--	19.94	555.77	20.68	555.03	22.13	553.58	22.17	553.54	17	I
10/9/01	11:37	--	--	14.49	560.97	--	--	20.42	555.29	21.12	554.59	22.47	553.24	22.50	553.21	15	I

Appendix 2C. Open-hole and discrete-zone manual water-level measurements in MW104R.—Continued

[Depth is depth to water, in feet. Elevation (elev.) is elevation of head, in feet above NAVD 88. "Z" means Zone in borehole. * means measurements referenced to 6-inch steel casing. All other measurements referenced to continuous multi-channel tubing. Numbers with parentheses indicates elevation of open zone in feet above North American Vertical Datum of 1988 (NAVD 88). Packer pressure in pounds force per square inch. ft, feet. NA, not applicable. NR, not recorded. I, inflated. L, low pressure. FZ, frozen. --, missing head data]

Date	Time	MW104R*		Z1* (557.5-539.5 ft)		Z1 (557.5-539.5 ft)		Z2 (537.5-503.5 ft)		Z3 (501.5-491.5 ft)		Z4 (489.5-472.5 ft)		Z5 (470.5-448.5 ft)		Packer pressure	Packer condition
		Depth	Elev.	Depth	Elev.	Depth	Elev.	Depth	Elev.	Depth	Elev.	Depth	Elev.	Depth	Elev.		
10/30/01	10:08	--	--	15.53	559.93	--	--	21.12	554.59	21.89	553.82	23.22	552.49	23.27	552.44	15	I
11/15/01	12:27	--	--	16.07	559.39	--	--	21.46	554.25	22.16	553.55	23.45	552.26	23.49	552.22	13	I
11/29/01	12:09	--	--	16.44	559.02	--	--	21.85	553.86	22.44	553.27	23.68	552.03	23.72	551.99	9	I
12/21/01	10:10	--	--	14.67	560.79	--	--	21.18	554.53	21.90	553.81	23.02	552.69	23.08	552.63	15	I
1/14/02	13:27	--	--	14.41	561.05	--	--	20.74	554.97	21.50	554.21	22.50	553.21	22.54	553.17	14	I
1/30/02	10:50	--	--	13.08	562.38	--	--	20.10	555.61	20.73	554.98	21.78	553.93	21.80	553.91	8	I
2/7/02	11:28	--	--	12.81	562.65	--	--	19.60	556.11	20.31	555.40	21.39	554.32	21.41	554.30	NR	FZ
2/28/02	13:31	--	--	12.66	562.80	--	--	19.25	556.46	20.01	555.70	21.11	554.60	21.11	554.60	11	I
3/28/02	11:28	--	--	9.12	566.34	--	--	16.85	558.86	17.73	557.98	18.99	556.72	19.04	556.67	18	I
4/16/02	11:40	--	--	9.25	566.21	--	--	15.97	559.74	16.96	558.75	18.32	557.39	18.36	557.35	14	I
5/7/02	12:50	--	--	8.48	566.98	--	--	15.46	560.25	16.50	559.21	17.98	557.73	18.03	557.68	21	I
6/18/02	14:45	--	--	7.88	567.58	--	--	13.98	561.73	15.30	560.41	17.23	558.48	17.30	558.41	20	I
7/10/02	14:48	--	--	11.27	564.19	--	--	16.28	559.43	17.50	558.21	--	--	19.52	556.19	18	I
7/11/02	10:40	--	--	11.34	564.12	--	--	16.26	559.45	17.50	558.21	--	--	19.50	556.21	16	I
7/12/02	10:34	--	--	11.55	563.91	--	--	16.36	559.35	17.55	558.16	--	--	19.52	556.19	16	I
8/21/02	14:39	--	--	14.28	561.18	--	--	19.38	556.33	20.56	555.15	22.51	553.20	22.54	553.17	14	I
9/17/02	14:40	--	--	15.37	560.09	--	--	20.40	555.31	21.55	554.16	23.22	552.49	23.27	552.44	20	I
10/28/02	12:48	--	--	12.87	562.59	--	--	19.71	556.00	20.72	554.99	22.29	553.42	22.34	553.37	18	I
11/26/02	13:30	--	--	9.75	565.71	--	--	17.04	558.67	18.06	557.65	19.46	556.25	19.50	556.21	18	I

Appendix 2D. Open-hole and discrete-zone manual water-level measurements in MW105R.

[Depth is depth to water, in feet. Elevation (elev.) is elevation of head, in feet above NAVD 88. * means measurements referenced to 6-inch steel casing. All other measurements referenced to continuous multi-channel tubing. Numbers with parentheses indicates elevation of open zone in feet above North American Vertical Datum of 1988 (NAVD 88). Packer pressure in pounds force per square inch. ft, feet. NA, not applicable. NR, not recorded. I, inflated. FZ, frozen. --, missing head data]

Date	Time	MW105R*		Z1* (553.8-511.3 ft)		Z1 (553.8-511.3 ft)		Z2 (509.3-495.8 ft)		Z3 (493.8-485.8 ft)		Z4 (483.8-459.8 ft)		Z5 (457.8-438.8 ft)		Packer pressure	Packer condition
		Depth	Elev.	Depth	Elev.	Depth	Elev.	Depth	Elev.	Depth	Elev.	Depth	Elev.	Depth	Elev.		
8/9/00	12:00	11.59	554.20	--	--	--	--	--	--	--	--	--	--	--	--	NR	NA
8/10/00	12:30	--	--	20.03	545.76	--	--	10.37	555.54	11.12	554.79	11.97	553.94	11.97	553.94	14	I
8/14/00	8:00	--	--	11.17	--	--	--	10.71	555.20	11.39	554.52	12.21	553.70	12.21	553.70	15	I
8/25/00	13:00	--	--	11.65	554.14	--	--	11.20	554.71	11.84	554.07	12.80	553.11	12.80	553.11	9	I
8/28/00	9:02	--	--	11.89	553.90	--	--	11.45	554.46	12.12	553.79	12.99	552.92	12.99	552.92	9	I
9/6/00	15:30	--	--	12.29	553.50	--	--	11.80	554.11	12.44	553.47	13.43	552.48	13.40	552.51	14	I
9/18/00	8:35	--	--	12.21	553.58	12.2	553.62	11.90	554.01	12.45	553.46	13.75	552.16	13.76	552.15	18	I
9/20/00	10:25	--	--	11.51	554.28	--	--	11.08	554.83	11.67	554.24	13.33	552.58	13.37	552.54	18	I
9/28/00	12:00	--	--	--	--	--	--	11.60	554.31	12.16	553.75	13.94	551.97	13.76	552.15	NR	NR
9/29/00	11:28	--	--	11.85	553.94	11.80	554.02	11.58	554.33	12.13	553.78	13.89	552.02	14.12	551.79	NR	I
10/20/00	8:45	--	--	12.69	553.10	12.7	553.12	12.46	553.45	12.92	552.99	14.41	551.50	14.44	551.47	16	I
10/31/00	12:00	--	--	--	--	--	--	13.03	552.88	13.34	552.57	14.73	551.18	14.74	551.17	NR	NR
11/2/00	12:24	--	--	13.07	552.72	13.08	552.74	12.93	552.98	13.35	552.56	14.72	551.19	14.75	551.16	NR	NR
11/27/00	12:30	--	--	11.47	554.32	11.49	554.33	11.81	554.10	12.20	553.71	14.10	551.81	14.17	551.74	21	I
12/1/00	12:00	--	--	--	--	--	--	--	--	--	--	14.02	551.89	14.61	551.30	NR	NR
12/11/00	11:56	--	--	11.63	554.16	11.65	554.17	11.88	554.03	12.22	553.69	13.74	552.17	13.84	552.07	20	I
12/21/00	14:30	--	--	9.71	556.08	9.72	556.10	10.05	555.86	10.15	555.76	10.98	554.93	11.33	554.58	NR	FZ
1/3/01	11:55	--	--	10.37	555.42	10.38	555.44	10.54	555.37	10.69	555.22	11.51	554.40	11.90	554.01	NR	FZ
2/13/01	11:59	--	--	9.95	555.84	9.97	555.85	10.43	555.48	10.58	555.33	11.35	554.56	11.89	554.02	NR	FZ
2/28/01	12:00	--	--	--	--	--	--	9.93	555.98	10.07	555.84	10.82	555.09	11.42	554.49	NR	FZ
3/14/01	11:35	--	--	9.09	556.70	9.11	556.71	9.67	556.24	9.80	556.11	10.58	555.33	11.25	554.66	NR	FZ
3/19/01	7:50	--	--	--	--	8.53	557.29	--	--	--	--	9.79	556.12	10.24	555.67	NR	FZ
4/5/01	13:21	--	--	7.95	557.84	--	--	8.38	557.53	8.49	557.42	8.85	557.06	9.07	556.84	NR	NR
4/20/01	11:08	--	--	8.59	557.20	--	--	8.92	556.99	9.05	556.86	9.57	556.34	9.89	556.02	NR	NR
4/26/01	12:00	--	--	--	--	--	--	9.21	556.70	9.31	556.60	9.73	556.18	10.45	555.46	NR	NR
5/1/01	10:44	--	--	9.20	556.59	--	--	9.43	556.48	9.57	556.34	10.39	555.52	10.65	555.26	NR	NR
5/16/01	12:55	--	--	10.15	555.64	--	--	10.27	555.64	10.36	555.55	11.35	554.56	11.80	554.11	NR	NR
6/4/01	12:00	--	--	--	--	--	--	9.53	556.38	--	--	10.68	555.23	11.28	554.63	NR	NR
6/8/01	12:30	--	--	9.51	556.28	--	--	9.72	556.19	9.84	556.07	10.74	555.17	11.28	554.63	NR	NR
6/25/01	11:12	--	--	8.96	556.83	--	--	9.29	556.62	9.38	556.53	10.13	555.78	10.55	555.36	NR	NR
7/10/01	10:53	--	--	10.53	555.26	--	--	10.59	555.32	10.69	555.22	11.58	554.33	12.27	553.64	NR	NR
7/24/01	15:08	--	--	11.86	553.93	--	--	11.76	554.15	11.88	554.03	12.80	553.11	13.59	552.32	NR	NR
8/13/01	10:52	--	--	13.12	552.67	--	--	12.36	553.55	12.66	553.25	13.75	552.16	14.82	551.09	NR	NR
8/30/01	14:00	--	--	13.22	552.57	13.19	552.63	12.88	553.03	13.01	552.90	14.27	551.64	15.19	550.72	13	I

Appendix 2D. Open-hole and discrete-zone manual water-level measurements in MW105R.—Continued

[Depth is depth to water, in feet. Elevation (elev.) is elevation of head, in feet above NAVD 88. "Z" means Zone in borehole. * means: measurements referenced to 6-inch steel casing. All other measurements referenced to continuous multi-channel tubing. Numbers with parentheses indicates elevation of open zone in feet above North American Vertical Datum of 1988 (NAVD 88). Packer pressure in pounds force per square inch. ft, feet. NA, not applicable. NR, not recorded. I, inflated. FZ, frozen. --, missing head data]

Date	Time	MW105R*		Z1* (553.8-511.3 ft)		Z1 (553.8-511.3 ft)		Z2 (509.3-495.8 ft)		Z3 (493.8-485.8 ft)		Z4 (483.8-459.8 ft)		Z5 (457.8-438.8 ft)		Packer pressure	Packer condition
		Depth	Elev.	Depth	Elev.	Depth	Elev.	Depth	Elev.	Depth	Elev.	Depth	Elev.	Depth	Elev.		
9/11/01	14:12	--	--	14.18	551.61	--	--	13.78	552.13	14.29	551.52	15.77	550.14	15.85	550.06	13	I
9/27/01	12:30	--	--	13.26	552.53	--	--	13.24	552.67	13.68	552.23	15.48	550.43	15.70	550.21	11	I
10/2/01	13:42	--	--	--	--	--	--	--	--	--	--	--	--	--	--	10	I
10/9/01	11:51	--	--	14.02	551.77	--	--	13.88	552.03	14.29	551.52	15.79	550.12	16.07	549.84	17	I
10/30/01	10:21	--	--	14.95	550.84	--	--	14.77	551.14	15.11	550.30	16.16	549.75	16.69	549.22	16	I
11/15/01	10:28	--	--	15.25	550.54	--	--	15.07	550.84	15.38	550.53	16.32	549.59	16.84	549.07	15	I
11/29/01	10:07	--	--	15.45	550.34	--	--	15.28	550.63	15.51	550.40	16.47	549.44	17.00	548.91	14	I
12/21/01	11:00	--	--	14.11	551.68	--	--	14.25	551.66	14.35	551.56	15.72	550.19	16.38	549.53	14	I
1/16/02	12:12	--	--	13.20	552.59	--	--	13.44	552.47	13.56	552.35	14.94	550.97	15.70	550.21	13	I
2/7/02	9:55	--	--	12.12	553.67	--	--	12.48	553.43	12.70	553.21	14.15	551.76	14.83	551.08	13	I
2/21/02	12:25	--	--	11.90	553.89	--	--	12.20	553.71	12.41	553.50	13.86	552.05	14.54	551.37	7	I
3/28/02	11:57	--	--	9.49	556.30	--	--	10.20	555.71	10.37	555.34	11.67	554.24	12.48	553.43	6	I
4/18/02	10:51	--	--	9.74	556.05	--	--	10.30	555.61	10.65	555.26	11.50	554.41	11.70	554.21	15	I
5/7/02	13:42	--	--	9.41	556.38	--	--	10.09	555.82	10.42	555.49	11.87	554.04	12.57	553.34	17	I
5/30/02	15:00	--	--	9.28	556.51	--	--	9.71	556.20	10.18	555.73	10.71	555.20	10.94	554.97	17	I
6/18/02	11:55	--	--	9.41	556.38	--	--	9.72	556.19	10.24	555.67	10.91	555.00	11.27	554.64	16	I
7/10/02	11:00	--	--	11.78	554.01	--	--	11.61	554.30	11.98	553.93	12.96	552.95	13.52	552.39	15	I
8/21/02	12:34	--	--	14.58	551.21	--	--	14.27	551.64	14.70	551.21	15.58	550.33	16.16	549.75	10	I
9/24/02	15:20	--	--	15.35	550.44	--	--	14.88	551.03	15.45	550.46	16.45	549.46	16.88	549.03	17	I
10/28/02	14:07	--	--	13.61	552.18	--	--	13.48	552.43	13.97	551.94	15.47	550.44	16.11	549.80	16	I
11/26/02	14:21	--	--	10.79	555.00	--	--	11.02	554.89	11.52	554.39	12.62	553.29	13.33	552.58	15	I

Appendix 2E. Open-hole and discrete-zone manual water-level measurements in MW109R.

[Depth is depth to water, in feet. Elevation (elev.) is elevation of head, in feet above NAVD 88. "Z" means Zone in borehole. * means measurements referenced to 6-inch steel casing. All other measurements referenced to continuous multi-channel tubing. Numbers with parentheses indicates elevation of open zone in feet above North American Vertical Datum of 1988 (NAVD 88). Packer pressure in pounds force per square inch. ft, feet. NA, not applicable. NR, not recorded. I, inflated. L, low pressure. FZ, frozen. --, missing head data]

Date	Time	MW109R* Depth	MW109R* Elev.	Z1 (562.3-547.3 ft) Depth	Z1 Elev.	Z2 (545.3-533.3 ft) Depth	Z2 Elev.	Z3 (531.3-514.3 ft) Depth	Z3 Elev.	Z4 (512.3-489.3 ft) Depth	Z4 Elev.	Z5 (487.3-455.3 ft) Depth	Z5 Elev.	Packer pressure	Packer condition
5/4/00	11:10	9.91	572.50	--	--	--	--	--	--	--	--	--	--	NR	NA
5/4/00	15:30	--	--	--	--	9.40	573.37	10.55	572.22	10.39	572.38	10.31	572.46	25	I
5/5/00	10:30	--	--	9.79	572.62	9.79	572.98	10.80	571.97	10.60	572.17	10.54	572.23	10	I
5/5/00	13:00	--	--	7.97	574.44	10.91	571.86	11.21	571.56	11.14	571.63	11.15	571.62	25	I
5/5/00	15:15	--	--	7.97	574.44	11.08	571.69	10.92	571.85	11.10	571.67	11.15	571.62	26	I
5/9/00	11:15	--	--	7.99	574.42	11.89	570.88	12.19	570.58	12.11	570.66	12.10	570.67	15	I
5/11/00	11:25	--	--	7.95	574.46	11.87	570.90	12.15	570.62	12.08	570.69	--	--	25	I
5/15/00	9:50	--	--	8.20	574.21	12.17	570.60	12.44	570.33	12.36	570.41	12.35	570.42	24	I
5/24/00	10:35	--	--	8.25	574.16	12.04	570.73	12.29	570.48	12.22	570.55	12.21	570.56	22	I
6/5/00	10:55	--	--	9.42	572.99	13.07	569.70	13.30	569.47	13.24	569.53	13.23	569.54	22	I
6/12/00	11:50	--	--	8.97	573.44	12.62	570.15	12.85	569.92	12.79	569.98	12.77	570.00	21	I
6/15/00	10:45	--	--	9.25	573.16	12.81	569.96	13.04	569.73	12.98	569.79	12.96	569.81	21	I
6/20/00	13:20	--	--	9.93	572.48	13.42	569.35	13.63	569.14	13.57	569.20	13.56	569.21	21	I
6/27/00	7:40	--	--	11.09	571.32	14.33	568.44	14.53	568.24	14.48	568.29	14.46	568.31	20	I
7/7/00	8:40	--	--	12.22	570.19	15.05	567.72	15.21	567.56	15.16	567.61	15.15	567.62	20	I
7/21/00	10:00	--	--	14.75	567.66	16.82	565.95	16.95	565.82	16.90	565.87	16.90	565.87	20	I
8/1/00	13:37	--	--	15.57	566.84	17.42	565.35	17.52	565.25	17.51	565.26	17.49	565.28	19	I
8/8/00	10:50	--	--	16.22	566.19	17.97	564.80	18.07	564.70	18.03	564.74	18.03	564.74	19	I
8/28/00	10:15	--	--	17.87	564.54	19.53	563.24	19.62	563.15	19.58	563.19	19.58	563.19	18	I
9/20/00	10:45	--	--	35.00	547.41	20.63	562.14	20.72	562.05	20.68	562.09	20.69	562.08	17	I
10/20/00	8:15	--	--	19.57	562.84	21.59	561.18	21.69	561.08	21.65	561.12	21.65	561.12	16	I
11/2/00	12:44	--	--	20.63	561.78	22.15	560.62	22.25	560.52	22.22	560.55	22.21	560.56	16	I
11/27/00	13:45	--	--	21.23	561.18	22.62	560.15	22.72	560.05	22.69	560.08	22.68	560.09	15	I
12/11/00	12:50	--	--	20.67	561.74	22.79	559.98	22.88	559.89	22.82	559.95	22.83	559.94	17	I
12/18/00	11:45	--	--	19.65	562.76	21.90	560.87	22.02	560.75	21.97	560.80	22.00	560.77	17	I
12/21/00	10:25	--	--	17.40	565.01	20.45	562.32	20.58	562.19	20.53	562.24	20.55	562.22	NR	FZ
1/3/01	12:10	--	--	16.69	565.72	19.29	563.48	19.42	563.35	19.38	563.39	19.37	563.40	NR	FZ
1/30/01	12:00	--	--	--	--	19.75	563.02	20.08	562.69	20.20	562.57	20.14	562.63	NR	FZ
2/1/01	15:55	--	--	17.75	564.66	19.89	562.88	20.02	562.75	19.98	562.79	19.95	562.82	NR	FZ
2/13/01	9:55	--	--	16.96	565.45	18.89	563.88	19.02	563.75	18.93	563.84	18.96	563.81	NR	FZ
3/14/01	8:50	--	--	12.71	569.70	15.82	566.95	15.98	566.79'	15.92	566.85	15.92	566.85	NR	FZ
4/5/01	10:25	--	--	6.74	575.67	--	--	11.12	571.65	11.05	571.72	--	--	NR	NR
4/20/01	9:25	--	--	7.54	574.87	11.49	571.28	11.69	571.08	11.64	571.13	11.64	571.13	NR	NR
5/1/01	12:10	--	--	8.64	573.77	12.27	570.50	12.47	570.30	12.41	570.36	12.41	570.36	NR	NR

Appendix 2E. Open-hole and discrete-zone manual water-level measurements in MW109R.—Continued

[Depth is depth to water, in feet. Elevation (elev.) is elevation of head, in feet above NAVD 88. "Z" means Zone in borehole. * means measurements referenced to 6-inch steel casing. All other measurements referenced to continuous multi-channel tubing. Numbers with parentheses indicates elevation of open zone in feet above North American Vertical Datum of 1988 (NAVD 88). Packer pressure in pounds force per square inch. ft, feet. NA, not applicable. NR, not recorded. I, inflated. L, low pressure. FZ, frozen. --, missing head data]

Date	Time	MW109R*		Z1 (562.3-547.3 ft)		Z2 (545.3-533.3 ft)		Z3 (531.3-514.3 ft)		Z4 (512.3-489.3 ft)		Z5 (487.3-455.3 ft)		Packer pressure	Packer condition
		Depth	Elev.	Depth	Elev.	Depth	Elev.	Depth	Elev.	Depth	Elev.	Depth	Elev.		
5/16/01	10:30	--	--	10.66	571.75	13.92	568.85	14.13	568.64	14.03	568.74	14.04	568.73	NR	NR
6/4/01	8:50	--	--	10.85	571.56	13.79	568.98	13.94	568.83	13.90	568.87	13.92	568.85	NR	NR
6/25/01	11:53	--	--	10.17	572.24	13.61	569.16	13.80	568.97	13.75	569.02	13.72	569.05	NR	NR
7/10/01	9:10	--	--	13.07	569.34	15.55	567.22	15.68	567.09	15.66	567.11	15.68	567.09	NR	NR
7/24/01	14:52	--	--	15.74	566.67	17.58	565.19	17.68	565.09	17.61	565.16	17.60	565.17	NR	NR
8/13/01	8:53	--	--	18.39	564.02	19.76	563.01	19.84	562.93	19.80	562.97	19.80	562.97	NR	NR
9/11/01	13:40	--	--	20.08	562.33	21.56	561.21	21.54	561.23	21.55	561.22	21.54	561.23	7	I
9/20/01	9:52	--	--	--	--	--	--	--	--	--	--	--	--	17	I
9/27/01	9:54	--	--	20.94	561.47	22.15	560.62	22.17	560.60	22.17	560.60	22.18	560.59	17	I
10/9/01	9:50	--	--	21.55	560.86	22.66	560.11	22.65	560.12	22.66	560.11	22.65	560.12	16	I
10/30/01	8:45	--	--	22.48	559.93	23.53	559.24	23.54	559.23	23.52	559.25	23.50	559.27	15	I
11/15/01	10:42	--	--	23.08	559.33	24.02	558.75	24.02	558.75	24.06	558.71	24.01	558.76	15	I
11/29/01	9:42	--	--	23.57	558.84	24.46	558.31	24.46	558.31	24.47	558.30	24.48	558.29	14	I
12/6/01	14:05	--	--	23.71	558.70	--	--	24.59	558.18	24.58	558.19	24.59	558.18	13	I
12/13/01	10:00	--	--	23.93	558.48	24.72	558.05	24.72	558.05	--	--	24.74	558.03	13	I
1/14/02	10:17	--	--	23.93	558.48	24.72	558.05	24.74	558.03	24.75	558.02	--	--	NR	FZ
1/31/02	10:45	--	--	23.60	558.81	24.32	558.45	24.33	558.44	24.28	558.49	--	--	NR	FZ
2/21/02	11:31	--	--	23.06	559.35	24.00	558.77	24.02	558.75	24.00	558.77	24.00	558.77	0	L
3/26/02	10:12	--	--	18.57	563.84	21.74	561.03	21.85	560.92	21.78	560.99	21.78	560.99	13	I
4/18/02	9:52	--	--	16.76	565.65	19.35	563.42	19.48	563.29	19.40	563.37	--	--	12	I
5/7/02	9:35	--	--	16.14	566.55	18.49	564.28	18.60	564.17	18.57	564.20	--	--	14	I
6/18/02	12:20	--	--	10.83	571.86	14.39	568.38	14.55	568.22	14.49	568.28	14.48	568.29	14	I
8/22/02	13:30	--	--	19.19	563.50	20.90	561.87	20.94	561.83	20.89	561.88	20.91	561.86	9	I
9/26/02	8:40	--	--	21.67	561.02	22.77	560.00	22.80	559.97	22.78	559.99	22.80	559.97	20	I
10/28/02	13:24	--	--	22.35	560.34	23.17	559.60	--	--	23.20	559.57	--	--	18.5	I
11/26/02	13:58	--	--	20.53	562.16	21.57	561.20	21.59	561.18	21.57	561.20	--	--	18	I

Appendix 2F. Open-hole and discrete-zone manual water-level measurements in MW121R.

[Depth is depth to water, in feet. Elevation (elev.) is elevation of head, in feet above NAVD 88. "Z" means Zone in borehole. * means measurements referenced to 6-inch steel casing. All other measurements referenced to continuous multi-channel tubing. Numbers with parentheses indicates elevation of open zone in feet above North American Vertical Datum of 1988 (NAVD 88). Packer pressure in pounds force per square inch. ft, feet. NA, not applicable. NR, not recorded. I, inflated. L, low pressure. FZ, frozen. --, missing head data]

Date	Time	MW121R*		Z1* (577.8-533.3 ft)		Z1 (577.8-533.3 ft)		Z2 (531.3-461.8 ft)		Packer pressure	Packer condition
		Depth	Elev.	Depth	Elev.	Depth	Elev.	Depth	Elev.		
4/18/00	12:00	20.00	568.82	--	--	19.19	570.01	20.88	568.32	30	I
4/20/00	12:00	--	--	15.45	573.37	15.87	573.33	21.29	567.91	21	I
4/29/00	14:45	--	--	13.61	573.71	13.65	575.55	20.33	568.87	0	L
5/5/00	15:40	--	--	15.11	573.71	15.50	573.70	21.00	568.20	16	I
5/9/00	12:50	--	--	15.75	573.07	16.13	573.07	21.27	567.93	16	I
5/11/00	11:50	--	--	15.92	572.90	16.30	572.90	21.18	568.02	14	I
5/15/00	10:25	--	--	16.19	572.63	16.56	572.64	21.46	567.74	22	I
5/24/00	11:10	--	--	15.56	573.26	15.93	573.27	21.14	568.06	20	I
6/5/00	11:20	--	--	17.08	571.74	17.43	571.77	21.96	567.24	16	I
6/12/00	12:10	--	--	15.78	573.04	16.13	573.07	21.40	567.80	16	I
6/15/00	9:40	--	--	16.22	572.60	16.57	572.63	21.51	567.69	30	I
6/20/00	13:15	--	--	17.47	571.35	17.84	571.36	21.97	567.23	24	I
6/26/00	12:00	--	--	--	--	18.83	570.37	21.02	568.18	NR	NR
6/27/00	8:10	--	--	18.74	570.08	19.12	570.08	22.66	566.54	22	I
7/7/00	9:05	--	--	19.21	569.61	19.59	569.61	22.87	566.33	22	I
7/21/00	10:15	--	--	20.71	568.11	21.08	568.12	23.51	565.69	20	I
8/1/00	13:55	--	--	20.65	568.17	21.02	568.18	23.03	566.17	20	I
8/8/00	10:00	--	--	21.01	567.81	21.39	567.81	23.70	565.50	20	I
8/17/00	12:00	--	--	--	--	21.18	568.02	23.44	565.76	NR	NR
8/28/00	9:55	--	--	22.67	566.15	23.06	566.14	25.05	564.15	20	I
9/6/00	14:15	--	--	23.12	565.70	23.45	565.75	25.37	563.83	19	I
9/20/00	10:15	--	--	23.29	565.53	23.65	565.55	25.01	564.19	18	I
9/28/00	12:00	--	--	--	--	23.40	565.80	25.30	563.90	NR	NR
9/29/00	11:00	--	--	23.06	565.76	23.34	565.86	25.23	563.97	18	I
10/20/00	8:35	--	--	24.10	564.72	24.46	564.74	26.13	563.07	16	I
10/31/00	12:00	--	--	--	--	24.36	564.84	--	--	NR	NR
11/2/00	12:01	--	--	24.35	564.47	24.75	564.45	27.42	561.78	16	I
11/27/00	13:09	--	--	23.26	565.56	23.64	565.56	24.91	564.29	10	I
12/11/00	12:32	--	--	22.58	566.24	22.95	566.25	24.51	564.69	8	I
12/18/00	12:20	--	--	21.02	567.80	21.37	567.83	25.98	563.22	13	I
12/21/00	14:11	--	--	15.94	572.88	16.32	572.88	21.22	567.98	5	L
1/3/01	11:28	--	--	17.22	571.60	17.57	571.63	21.75	567.45	NR	FZ
2/13/01	12:27	--	--	17.18	571.64	17.56	571.64	21.57	567.63	NR	FZ

Appendix 2F. Open-hole and discrete-zone manual water-level measurements in MW121R.—Continued

[Depth is depth to water, in feet. Elevation (elev.) is elevation of head, in feet above NAVD 88. "Z" means Zone in borehole. * means measurements referenced to 6-inch steel casing. All other measurements referenced to continuous multi-channel tubing. Numbers with parentheses indicates elevation of open zone in feet above North American Vertical Datum of 1938 (NAVD 88). Packer pressure in pounds force per square inch. ft, feet. NA, not applicable. NR, not recorded. I, inflated. L, low pressure. FZ, frozen. --, missing head data]

Date	Time	MW121R*		Z1* (577.8-533.3 ft)		Z1 (577.8-533.3 ft)		Z2 (531.3-461.8 ft)		Packer pressure	Packer condition
		Depth	Elev.	Depth	Elev.	Depth	Elev.	Depth	Elev.		
2/28/01	12:00	--	--	--	--	16.87	572.33	21.15	568.05	NR	FZ
3/14/01	11:50	--	--	16.86	571.96	17.21	571.99	20.94	568.26	NR	FZ
4/5/01	14:09	--	--	13.28	575.54	13.65	575.55	19.45	569.75	15	I
4/20/01	11:22	--	--	15.19	573.63	15.52	573.68	20.40	568.80	NR	NR
5/1/01	10:20	--	--	16.63	572.19	17.01	572.19	21.02	568.18	NR	NR
5/16/01	14:25	--	--	18.57	570.25	18.95	570.25	22.33	566.87	NR	NR
6/4/01	12:00	--	--	--	--	19.09	570.11	21.47	567.73	NR	NR
6/8/01	11:37	--	--	17.71	571.11	18.07	571.13	21.78	567.42	NR	NR
6/25/01	11:27	--	--	16.34	572.48	16.69	572.51	21.12	568.08	NR	NR
7/10/01	11:13	--	--	19.64	569.18	20.00	569.20	22.98	566.22	NR	NR
7/24/01	15:28	--	--	21.94	566.88	22.31	566.89	24.80	564.40	NR	NR
8/13/01	11:15	--	--	24.26	564.56	24.65	564.55	26.39	562.81	NR	NR
9/11/01	14:33	--	--	25.70	563.12	25.98	563.22	27.62	561.58	11	I
9/27/01	12:58	--	--	25.59	563.23	25.94	563.26	27.28	561.92	18	I
10/2/01	13:59	--	--	--	--	--	--	--	--	18	I
10/9/01	12:06	--	--	26.01	562.81	26.41	562.79	27.84	561.36	15	I
10/30/01	10:33	--	--	26.91	561.91	27.33	561.87	28.76	560.44	12	I
11/15/01	9:46	--	--	27.35	561.47	27.76	561.44	29.13	560.07	12	I
11/29/01	10:22	--	--	27.70	561.12	28.13	561.07	29.48	559.72	14	I
12/27/01	10:53	--	--	26.76	562.06	27.19	562.01	27.77	561.43	13	I
1/16/02	12:45	--	--	26.11	562.71	26.52	562.68	27.16	562.04	11	I
2/7/02	9:15	--	--	24.72	564.10	25.10	564.10	26.09	563.11	13	I
2/21/02	13:06	--	--	24.08	564.74	24.48	564.72	25.49	563.71	0	L
3/28/02	13:18	--	--	18.22	570.60	18.66	570.54	22.21	566.99	17	I
4/16/02	9:50	--	--	17.17	571.65	17.45	571.75	21.68	567.52	9	I
5/7/02	14:52	--	--	16.04	572.78	16.45	572.75	21.36	567.84	26	I
5/30/02	13:50	--	--	16.47	572.35	16.87	572.33	21.28	567.92	22	I
6/18/02	11:05	--	--	16.86	571.96	17.25	571.95	21.34	567.86	20	I
8/22/02	12:54	--	--	25.11	563.71	25.53	563.67	27.52	561.68	12	I
9/26/02	11:46	--	--	26.48	562.34	26.89	562.31	28.53	560.67	23	I
10/28/02	15:02	--	--	25.77	563.05	26.19	563.01	27.13	562.07	20	I

Appendix 2G. Open-hole and discrete-zone manual water-level measurements in MW122R.

[Depth is depth to water, in feet. Elevation (elev.) is elevation of head, in feet above NAVD 88. "Z" means Zone in borehole. * means measurements referenced to 6-inch steel casing. All other measurements referenced to continuous multi-channel tubing. Numbers with parentheses indicates elevation of open zone in feet above North American Vertical Datum of 1988 (NAVD 88). Packer pressure in pounds force per square inch. ft, feet. NA, not applicable. NR, not recorded. I, inflated. L, low pressure. FZ, frozen. --, missing head data]

Date	Time	MW122R*		Z1* (569.9-522.9 ft)		Z1 (569.9-522.9 ft)		Z2 (520.9-511.4 ft)		Z3 (509.4-454.4 ft)		Packer pressure	Packer condition
		Depth	Elev.	Depth	Elev.	Depth	Elev.	Depth	Elev.	Depth	Elev.		
7/27/00	16:55	14.25	567.15	--	--	--	--	--	--	--	--	NR	NA
7/27/00	18:40	--	--	13.36	568.04	13.71	568.05	14.37	567.39	13.73	568.03	24	I
8/1/00	12:58	--	--	13.43	--	13.80	567.96	15.87	565.89	13.80	567.96	12	I
8/8/00	10:10	--	--	14.21	567.19	14.57	567.19	14.93	566.83	14.57	567.19	13	I
8/17/00	12:00	--	--	--	--	14.43	567.33	14.47	567.29	14.43	567.33	NR	NR
8/28/00	11:08	--	--	17.8	-999.00	16.13	565.63	16.44	565.32	16.13	565.63	6	I
9/6/00	15:45	--	--	--	--	16.62	565.14	16.59	565.17	16.62	565.14	15	I
9/20/00	10:00	--	--	15.11	566.29	15.43	566.33	15.58	566.18	15.44	566.32	22	I
9/28/00	12:00	--	--	--	--	16.02	565.74	16.22	565.54	16.01	565.75	NR	NR
9/29/00	10:50	--	--	15.75	565.65	16.07	565.69	16.26	565.50	16.08	565.68	22	I
10/20/00	9:15	--	--	17.11	564.29	17.43	564.33	17.56	564.20	17.44	564.32	20	I
11/2/00	11:48	--	--	17.53	563.87	17.86	563.90	17.99	563.77	17.86	563.90	20	I
11/27/00	10:54	--	--	15.53	565.87	15.87	565.89	16.07	565.69	15.87	565.89	8	I
12/1/00	12:00	--	--	--	--	15.62	566.14	15.43	566.33	15.66	566.10	NR	NR
12/11/00	11:05	--	--	15.37	566.03	15.61	566.15	15.93	565.83	15.62	566.14	16	I
12/21/00	11:33	--	--	10.36	571.04	10.68	571.08	12.46	569.30	10.67	571.09	NR	FZ
1/3/01	10:45	--	--	11.03	570.37	11.35	570.41	13.42	568.34	11.35	570.41	NR	FZ
2/1/01	16:42	--	--	11.07	570.33	11.37	570.39	12.69	569.07	11.39	570.37	NR	FZ
2/13/01	11:45	--	--	10.88	570.52	11.20	570.56	12.70	569.06	11.20	570.56	NR	FZ
3/14/01	10:35	--	--	9.75	571.65	10.07	571.69	11.97	569.79	10.07	571.69	NR	FZ
3/16/01	11:40	--	--	9.82	571.58	--	--	11.88	569.88	--	--	NR	FZ
4/5/01	12:16	--	--	7.55	573.85	7.85	573.91	10.22	571.54	7.84	573.92	17	I
4/20/01	10:21	--	--	8.90	572.50	9.22	572.54	11.50	570.26	9.22	572.54	NR	NR
5/1/01	11:48	--	--	9.93	571.47	10.25	571.51	12.35	569.41	10.26	571.50	NR	NR
5/16/01	11:40	--	--	11.98	569.42	12.31	569.45	13.82	567.94	12.35	569.41	NR	NR
6/4/01	10:37	--	--	10.04	571.36	10.35	571.41	12.47	569.29	10.37	571.39	NR	NR
6/25/01	10:33	--	--	10.18	571.22	10.49	571.27	12.54	569.22	10.52	571.24	NR	NR
7/10/01	10:16	--	--	13.04	568.36	13.38	568.38	14.63	567.13	13.39	568.37	NR	NR
7/24/01	10:59	--	--	14.98	566.42	15.31	566.45	16.33	565.43	15.28	566.48	NR	NR
8/13/01	9:56	--	--	17.55	563.85	17.89	563.87	17.90	563.86	17.91	563.85	NR	NR
8/30/01	12:00	--	--	17.93	563.47	--	--	18.25	563.51	17.95	563.81	NR	NR
9/11/01	11:25	--	--	18.45	562.95	--	--	19.02	562.74	18.68	563.08	25	I
9/27/01	11:45	--	--	17.66	563.74	17.97	563.79	18.33	563.43	17.94	563.82	23	I
10/9/01	11:07	--	--	18.20	563.20	18.52	563.24	18.94	562.82	18.55	563.21	18	I

Appendix 2G. Open-hole and discrete-zone manual water-level measurements in MW122R.—Continued

[Depth is depth to water, in feet. Elevation (elev.) is elevation of head, in feet above NAVD 88. "Z" means Zone in borehole. * means measurements referenced to 6-inch steel casing. All other measurements referenced to continuous multi-channel tubing. Numbers with parentheses indicates elevation of open zone in feet above North American Vertical Datum of 1988 (NAVD 88). Packer pressure in pounds force per square inch. ft, feet. NA, not applicable. NR, not recorded. I, inflated. L, low pressure. FZ, frozen. --, missing head data]

Date	Time	MW122R*		Z1* (569.9-522.9 ft)		Z1 (569.9-522.9 ft)		Z2 (520.9-511.4 ft)		Z3 (509.4-454.4 ft)		Packer pressure	Packer condition
		Depth	Elev.	Depth	Elev.	Depth	Elev.	Depth	Elev.	Depth	Elev.		
10/11/01	13:52	--	--	18.29	563.11	18.60	563.16	19.01	562.75	18.61	563.15	17	I
10/30/01	9:48	--	--	19.08	562.32	19.42	562.34	19.90	561.86	19.43	562.33	17	I
11/15/01	11:55	--	--	19.59	561.81	19.93	561.83	20.57	561.19	19.94	561.82	16	I
11/27/00	10:54	--	--	--	--	15.87	565.89	16.07	565.69	15.87	565.89	8	I
11/29/01	11:58	--	--	19.92	561.48	20.25	561.51	20.67	561.09	20.27	561.49	14	I
12/11/00	11:05	--	--	15.37	566.03	15.61	566.15	15.93	565.83	15.62	566.14	16	I
12/13/01	11:32	--	--	19.91	561.49	20.23	561.53	20.65	561.11	20.23	561.53	13	I
1/14/02	12:37	--	--	17.24	564.16	17.56	564.20	17.97	563.79	17.55	564.21	13	I
1/30/02	10:20	--	--	--	--	16.78	564.98	17.38	564.38	16.78	564.98	NR	NR
2/7/02	10:52	--	--	16.04	565.36	16.35	565.41	16.80	564.96	16.32	565.44	15	I
2/28/02	12:03	--	--	15.32	566.08	15.64	566.12	16.41	565.35	15.63	566.13	NR	FZ
3/6/02	9:33	--	--	14.08	567.32	14.46	567.30	15.25	566.51	14.43	567.33	NR	FZ
3/28/02	10:29	--	--	10.41	570.99	10.70	571.06	12.60	569.16	10.71	571.05	NR	FZ
4/16/02	10:42	--	--	10.33	571.07	10.57	571.19	13.17	568.59	10.56	571.2	0	L
5/2/02	12:05	--	--	--	--	--	--	--	--	--	--	0	L
5/7/02	12:08	--	--	9.84	571.56	10.15	571.61	12.56	569.20	10.15	571.61	17	I
6/18/02	13:59	--	--	10.08	571.32	10.41	571.35	10.70	--	10.42	571.34	11	I
7/10/02	13:55	--	--	--	--	14.73	567.03	--	--	14.75	567.01	20	I
7/12/02	9:32	--	--	15.20	566.20	15.48	566.28	15.87	565.89	15.48	566.28	17	I
8/21/02	13:52	--	--	18.08	563.32	18.44	563.32	18.87	562.89	18.43	563.33	13	I
9/24/02	14:11	--	--	18.84	562.56	19.19	562.57	19.62	562.14	19.20	562.56	22	I
10/28/02	12:10	--	--	16.95	564.45	17.29	564.47	17.82	563.94	17.29	564.47	17	I
11/19/02	14:35	--	--	13.78	567.62	14.11	567.65	14.86	566.90	14.11	567.65	11	I
11/26/02	13:00	--	--	13.92	567.48	14.27	567.49	14.92	566.84	14.27	567.49	11	I

Appendix 2H. Open-hole and discrete-zone manual water-level measurements in MW201R.

[Depth is depth to water, in feet. Elevation (elev.) is elevation of head, in feet above NAVD 88. "Z" means Zone in borehole. FLOW indicates head above measuring point. * means measurements referenced to 6-inch steel casing. All other measurements referenced to continuous multi-channel tubing. Numbers with parentheses indicates elevation of open zone in feet above North American Vertical Datum of 1988 (NAVD 88). Packer pressure in pounds force per square inch. ft, feet. NA, not applicable. NR, not recorded. I, inflated. L, low pressure. FZ, frozen. --, missing head data]

Date	Time	MW201R* Depth	MW201R* Elev.	MW201R* Flow	Z1* (539.0-505.2 ft) Depth	Z1* Elev.	Z1* Flow	Z1 (539.0-505.2 ft) Depth	Z1 Elev.	Z1 Flow	Z2 (503.2-485.2 ft) Depth	Z2 Elev.	Z2 Flow	Z3 (483.2-465.2 ft) Depth	Z3 Elev.	Z3 Flow	Z4 (463.2-428.6 ft) Depth	Z4 Elev.	Z4 Flow	Packer pressure	Packer condition
6/22/00	12:00	2.15	552.07		--	--		--	--		--	--		--	--		--	--		NA	NA
6/24/00	12:00	1.00	553.22		--	--		--	--		--	--		--	--		--	--		NA	NA
6/26/00	8:30	1.00	553.22		--	--		--	--		--	--		--	--		--	--		NA	NA
8/16/00	10:00	1.75	552.47		--	--		--	--		--	--		--	--		--	--		NA	NA
8/17/00	12:00	1.64	552.58		--	--		--	--		--	--		--	--		--	--		NA	NA
8/21/00	9:00	2.00	552.22		--	--		--	--		--	--		--	--		--	--		NA	NA
8/25/00	12:00	2.74	551.48		--	--		--	--		--	--		--	--		--	--		NA	NA
8/29/00	8:32	2.79	551.43		--	--		--	--		--	--		--	--		--	--		NA	NA
8/31/00	11:10	2.87	551.35		--	--		--	--		--	--		--	--		--	--		NA	NA
9/5/00	10:03	2.59	551.63		--	--		--	--		--	--		--	--		--	--		NA	NA
9/7/00	9:40	2.99	551.23		--	--		--	--		--	--		--	--		--	--		NA	NA
9/8/00	10:00	2.82	551.40		--	--		--	--		--	--		--	--		--	--		NA	NA
9/27/00	13:30	2.64	551.58		--	--		--	--		--	--		--	--		--	--		NA	NA
10/31/00	12:00	4.03	550.19		--	--		--	--		--	--		--	--		--	--		NA	NA
11/27/00	14:15	1.44	552.78		--	--		--	--		--	--		--	--		--	--		NA	NA
12/1/00	12:00	3.20	551.02		--	--		--	--		--	--		--	--		--	--		NA	NA
12/8/00	11:37	3.18	551.04		--	--		--	--		--	--		--	--		--	--		NA	NA
1/3/01	12:41	1.31	552.91		--	--		--	--		--	--		--	--		--	--		NA	NA
1/30/01	12:00	1.98	552.24		--	--		--	--		--	--		--	--		--	--		NA	NA
2/2/01	9:26	1.94	552.28		--	--		--	--		--	--		--	--		--	--		NA	NA
2/13/01	9:36	1.36	552.86		--	--		--	--		--	--		--	--		--	--		NA	NA
2/28/01	12:00	1.10	553.12		--	--		--	--		--	--		--	--		--	--		NA	NA
3/29/01	16:25	0.00	554.22	FLOW	--	--		--	--		--	--		--	--		--	--		NA	NA
3/29/01	16:25	0.00	554.22	FLOW	--	--		--	--		--	--		--	--		--	--		NA	NA
3/29/01	17:33	--	--		0.00	554.22	FLOW	0.32	554.22		0.00	554.68	FLOW	0.00	554.68	FLOW	0.00	554.68	FLOW	20	I
4/5/01	10:42	--	--		0.00	554.22	FLOW	0.32	554.22		0.00	554.68	FLOW	0.00	554.68	FLOW	0.00	554.68	FLOW	16	I
4/6/01	10:35	--	--		0.00	554.22	FLOW	0.32	554.22		0.00	554.68	FLOW	0.00	554.68	FLOW	0.00	554.68	FLOW	16	I
4/10/01	12:00	--	--		0.00	554.22	FLOW	0.32	554.22		0.00	554.68	FLOW	0.00	554.68	FLOW	0.00	554.68	FLOW	15	I
4/20/01	10:56	--	--		0.07	554.15		0.39	554.15		0.00	554.68	FLOW	0.00	554.68	FLOW	0.00	554.68	FLOW	12	I
4/24/01	13:34	--	--		0.43	553.79		0.75	553.79		0.00	554.68	FLOW	0.00	554.68	FLOW	0.30	554.38		NR	NR
5/1/01	12:36	--	--		0.90	553.32		1.21	553.33		0.00	554.68	FLOW	0.14	554.54	FLOW	1.26	553.42		10	I
5/16/01	13:13	--	--		1.95	552.27		2.23	552.31		1.13	553.55		1.30	553.38		1.68	553.00		8	I
6/8/01	12:02	--	--		1.55	552.67		1.85	552.69		0.48	554.20		0.75	553.93		3.90	550.78		16	I
7/10/01	11:50	--	--		2.37	551.85		2.72	551.82		1.49	553.19		1.66	553.02		4.22	550.46		12	I

Appendix 2H. Open-hole and discrete-zone manual water-level measurements in MW201R.—Continued

[Depth is depth to water, in feet. Elevation (elev.) is elevation of head, in feet above NAVD 88. "Z" means Zone in borehole. FLOW indicates head above measuring point. * means measurements referenced to 6-inch steel casing. All other measurements referenced to continuous multi-channel tubing. Numbers with parentheses indicates elevation of open zone in feet above North American Vertical Datum of 1988 (NAVD 88). Packer pressure in pounds force per square inch. ft, feet. NA, not applicable. NR, not recorded. I, inflated. L, low pressure. FZ, frozen. --, missing head data]

Date	Time	MW201R*			Z1* (539.0-505.2 ft)			Z1 (539.0-505.2 ft)			Z2 (503.2-465.2 ft)			Z3 (483.2-465.2 ft)			Z4 (463.2-428.6 ft)			Packer pressure	Packer condition
		Depth	Elev.	Flow	Depth	Elev.	Flow	Depth	Elev.	Flow	Depth	Elev.	Flow	Depth	Elev.	Flow	Depth	Elev.	Flow		
7/24/01	16:00	--	--		3.63	550.59		3.89	550.65		2.78	551.90		2.92	551.76		4.96	549.72		12	I
8/13/01	11:42	--	--		4.62	549.6		4.87	549.67		3.98	550.70		4.09	550.59		5.73	548.95		10	I
9/27/01	13:29	--	--		5.44	548.78		5.72	548.82		4.86	549.82		4.95	549.73		7.92	546.76		15	I
10/9/01	9:30	--	--		5.82	548.4		6.12	548.42		5.22	549.46		5.31	549.37		8.17	546.51		15	I
10/30/01	11:02	--	--		6.12	548.1		6.42	548.12		5.93	548.75		5.90	548.78		5.98	--		0	L
11/1/01	9:15	--	--		6.34	547.88		6.62	547.92		5.78	548.90		5.85	548.83		8.14	546.54		18	I
11/15/01	9:14	--	--		6.41	547.81		6.72	547.82		5.97	548.71		5.99	548.69		8.42	546.26		13	I
11/29/01	9:23	--	--		6.51	547.71		6.82	547.72		6.08	548.60		6.14	548.54		8.39	546.29		12	I
12/27/01	10:30	--	--		5.43	548.79		5.74	548.8		5.21	549.47		5.22	549.46		5.35	--		7	I
1/14/02	9:55	--	--		5.3	548.92		5.58	548.96		4.91	549.77		4.98	549.7		7.58	547.1		15	I
2/7/02	11:59	--	--		4.47	549.75		4.79	549.75		3.96	550.72		4.02	550.66		6.75	547.93		13	I
2/21/02	11:06	--	--		4.17	550.05		4.46	550.08		3.65	551.03		3.73	550.95		6.25	548.43		12	I
3/26/02	9:50	--	--		2.79	551.43		3.08	551.46		2.10	552.53		2.18	552.5		5.43	549.25		14	I
4/18/02	11:25	--	--		1.87	552.35		2.18	552.36		0.92	553.75		0.98	553.7		4.31	550.37		12	I
5/2/02	10:58	--	--		1.92	552.3		2.22	552.32		0.82	553.85		0.95	553.73		4.09	550.59		13	I
5/16/02	15:00	--	--		1.05	553.17		--	--		0.00	554.63	FLOW	0.00	554.68	FLOW	--	--		NR	NR
5/30/02	13:25	--	--		1.17	553.05		1.46	553.08		0.13	554.55		0.25	554.43		3.98	550.7		20	I
6/18/02	10:45	--	--		1.71	552.51		2.00	552.54		0.58	554.10		0.71	553.97		4.31	550.37		19	I
8/22/02	13:59	--	--		6.27	547.95		6.57	547.97		5.45	549.23		5.54	549.14		8.38	546.3		15	I
9/26/02	12:09	--	--		6.9	547.32		7.19	547.35		6.18	548.50		6.26	548.42		9.12	545.56		17	I
10/28/02	15:33	--	--		5.96	548.26		6.26	548.28		5.32	549.35		5.41	549.27		8.23	546.45		16	I

Appendix 2I. Open-hole and discrete-zone manual water-level measurements in MW202R.

[Depth is depth to water, in feet. Elevation (elev.) is elevation of head, in feet above NAVD 88. "Z" means Zone in borehole. * means measurements referenced to 6-inch steel casing. All other measurements referenced to continuous multi-channel tubing. Numbers with parentheses indicates elevation of open zone in feet above North American Vertical Datum of 1988 (NAVD 88). Packer pressure in pounds force per square inch. ft, feet. NA, not applicable. NR, not recorded. I, inflated. FZ, frozen. --, missing head data]

Date	Time	MW202R* Depth	MW202R* Elev.	Z1* (568.8-545.3 ft) Depth	Z1* Elev.	Z1 (568.8-545.3 ft) Depth	Z1 Elev.	Z2 (543.3-483.3 ft) Depth	Z2 Elev.	Z3 (481.3-455.4 ft) Depth	Z3 Elev.	Packer pressure	Packer condition
6/20/00	12:00	11.49	570.84	--	--	--	--	--	--	--	--	NA	NA
8/3/00	20:08	14.85	567.48	--	--	--	--	--	--	--	--	NA	NA
8/4/00	8:25	14.29	568.04	--	--	--	--	--	--	--	--	NA	NA
8/4/00	13:35	14.55	567.78	--	--	--	--	--	--	--	--	NA	NA
8/11/00	12:00	14.90	567.43	--	--	--	--	--	--	--	--	NA	NA
8/16/00	9:29	14.78	567.55	--	--	--	--	--	--	--	--	NA	NA
8/18/00	9:00	14.30	568.03	--	--	--	--	--	--	--	--	NA	NA
8/22/00	12:00	14.95	567.38	--	--	--	--	--	--	--	--	NA	NA
8/28/00	8:24	15.42	566.91	--	--	--	--	--	--	--	--	NA	NA
9/6/00	8:24	15.39	566.94	--	--	--	--	--	--	--	--	NA	NA
9/6/00	12:33	15.11	567.22	--	--	--	--	--	--	--	--	NA	NA
9/7/00	16:48	15.47	566.86	--	--	--	--	--	--	--	--	NA	NA
10/4/00	9:45	15.81	566.52	--	--	--	--	--	--	--	--	NA	NA
10/5/00	10:05	16.15	566.18	--	--	--	--	--	--	--	--	NA	NA
10/31/00	12:00	17.72	564.61	--	--	--	--	--	--	--	--	NA	NA
11/27/00	12:35	16.82	565.51	--	--	--	--	--	--	--	--	NA	NA
12/1/00	12:00	16.42	565.91	--	--	--	--	--	--	--	--	NA	NA
12/8/00	11:19	15.57	566.76	--	--	--	--	--	--	--	--	NA	NA
1/3/01	11:43	11.89	570.44	--	--	--	--	--	--	--	--	NA	NA
1/30/01	12:00	13.09	569.24	--	--	--	--	--	--	--	--	NA	NA
2/2/01	9:05	12.58	569.75	--	--	--	--	--	--	--	--	NA	NA
2/13/01	12:20	12.57	569.76	--	--	--	--	--	--	--	--	NA	NA
2/28/01	12:00	12.17	570.16	--	--	--	--	--	--	--	--	NA	NA
3/29/01	11:32	8.32	574.01	--	--	--	--	--	--	--	--	NA	NA
3/29/01	12:49	--	--	4.41	577.92	4.93	577.91	6.03	576.81	12.12	570.72	NA	NA
4/5/01	14:25	--	--	4.57	577.76	5.09	577.75	6.91	575.93	13.16	569.68	21	I
4/20/01	11:31	--	--	6.48	575.85	6.98	575.86	8.89	573.95	14.25	568.59	12	I
4/26/01	12:00	--	--	--	--	8.37	574.47	8.91	573.93	15.19	567.65	7	I
5/1/01	7:26	--	--	8.07	574.26	8.58	574.26	10.61	572.23	14.82	568.02	NR	NR
5/16/01	14:17	--	--	10.44	571.89	10.96	571.88	13.37	569.47	16.26	566.58	NR	NR
6/4/01	12:00	--	--	--	--	10.73	572.11	--	--	--	--	NR	NR
6/5/01	12:30	--	--	9.98	572.35	10.49	572.35	12.93	569.91	15.59	567.25	NR	NR
6/25/01	11:35	--	--	8.53	573.80	9.02	573.82	12.60	570.24	15.03	567.81	NR	NR
7/10/01	11:22	--	--	11.50	570.83	12.03	570.81	15.17	567.67	17.10	565.74	NR	NR

Appendix 2I. Open-hole and discrete-zone manual water-level measurements in MW202R.—Continued

[Depth is depth to water, in feet. Elevation (elev.) is elevation of head, in feet above NAVD 88. "Z" means Zone in borehole. * means measurements referenced to 6-inch steel casing. All other measurements referenced to continuous multi-channel tubing. Numbers with parentheses indicates elevation of open zone in feet above North American Vertical Datum of 1988 (NAVD 88). Packer pressure in pounds force per square inch. ft, feet. NA, not applicable. NR, not recorded. I, inflated. FZ, frozen. --, missing head data]

Date	Time	MW202R* Depth	MW202R* Elev.	Z1* (568.8-545.3 ft) Depth	Z1* (568.8-545.3 ft) Elev.	Z1 (568.8-545.3 ft) Depth	Z1 (568.8-545.3 ft) Elev.	Z2 (543.3-483.3 ft) Depth	Z2 (543.3-483.3 ft) Elev.	Z3 (481.3-455.4 ft) Depth	Z3 (481.3-455.4 ft) Elev.	Packer pressure	Packer condition
7/24/01	15:36	--	--	13.42	568.91	13.93	568.91	16.87	565.97	19.82	563.32	NR	NR
8/13/01	11:24	--	--	15.52	566.81	16.06	566.78	18.75	564.09	20.67	562.17	NR	NR
9/11/01	14:52	--	--	16.99	565.34	17.43	565.41	18.32	564.52	21.79	561.35	17	I
9/27/01	13:09	--	--	17.49	564.84	18.02	564.82	18.65	564.19	21.71	561.13	14	I
10/9/01	12:15	--	--	17.81	564.52	18.33	564.51	19.1	563.74	22.27	560.57	14	I
10/30/01	10:41	--	--	18.43	563.9	18.96	563.88	19.67	563.17	23.15	559.69	11	I
11/15/01	10:05	--	--	18.82	563.51	19.37	563.47	20.04	562.80	23.46	559.38	13	I
11/29/01	10:35	--	--	19.16	563.17	19.7	563.14	20.33	562.51	23.77	559.07	11	I
12/27/01	11:03	--	--	19.11	563.22	19.67	563.17	19.77	563.07	22.25	560.59	14	I
1/16/02	12:59	--	--	18.85	563.48	19.38	563.46	19.49	563.35	21.69	561.15	12	I
2/7/02	9:28	--	--	17.77	564.56	18.3	564.54	18.46	564.38	20.55	562.29	15	I
2/21/02	13:28	--	--	16.91	565.42	17.43	565.41	17.57	565.27	19.9	562.94	14	I
2/28/02	14:15	--	--	16.2	566.13	16.73	566.11	17.71	565.13	19.76	563.08	8	I
3/28/02	13:30	--	--	9.86	572.47	10.36	572.48	12.85	569.99	16.56	566.18	15	I
4/18/02	11:06	--	--	8.55	573.78	9.08	573.76	11.04	571.80	15.85	566.99	12	I
5/7/02	15:01	--	--	7.74	574.59	8.23	574.61	10.38	572.46	15.5	567.34	23	I
5/30/02	14:02	--	--	7.68	574.65	8.18	574.66	10.11	572.73	15.55	567.29	21	I
6/18/02	11:15	--	--	8.20	574.13	8.7	574.14	10.57	572.27	15.55	567.29	19	I
6/18/02	11:15	--	--	8.20	574.13	8.7	574.14	10.57	572.27	15.55	567.29	19	I
8/22/02	13:03	--	--	15.88	566.45	16.42	566.42	17.57	565.27	21.8	561.04	13	I
9/26/02	11:56	--	--	17.68	564.65	18.24	564.6	19.03	563.81	22.75	560.09	19	I
10/28/02	15:15	--	--	18.16	-18.16	18.72	564.12	19.00	563.84	21.56	561.28	16.5	I

Appendix 2J. Open-hole and discrete-zone manual water-level measurements in MW203R.

[Depth is depth to water, in feet. Elevation (elev.) is elevation of head, in feet above NAVD 88. "Z" means Zone in borehole. * means measurements referenced to 6-inch steel casing. All other measurements referenced to continuous multi-channel tubing. Numbers with parentheses indicates elevation of open zone in feet above North American Vertical Datum of 1988 (NAVD 88). Packer pressure in pounds force per square inch. ft, feet. NA, not applicable. NR, not recorded. I, inflated. L, low pressure; FZ, frozen. --, missing head data]

Date	Time	MW203R* Depth	MW203R* Elev.	Z1* (562.5-533.9 ft) Depth	Z1* (562.5-533.9 ft) Elev.	Z1 (562.5-533.9 ft) Depth	Z1 (562.5-533.9 ft) Elev.	Z2 (531.9-488.9 ft) Depth	Z2 (531.9-488.9 ft) Elev.	Z3 (486.9-451.9 ft) Depth	Z3 (486.9-451.9 ft) Elev.	Packer pressure	Packer condition
6/22/00	9:35	10.47	566.43	--	--	--	--	--	--	--	--	NA	NA
6/23/00	9:00	10.56	566.34	--	--	--	--	--	--	--	--	NA	NA
8/15/00	12:20	13.06	563.84	--	--	--	--	--	--	--	--	NA	NA
8/23/00	12:00	13.65	563.25	--	--	--	--	--	--	--	--	NA	NA
8/28/00	12:15	14.06	562.84	--	--	--	--	--	--	--	--	NA	NA
9/11/00	11:11	14.37	562.53	--	--	--	--	--	--	--	--	NA	NA
9/12/00	7:21	14.72	562.18	--	--	--	--	--	--	--	--	NA	NA
9/12/00	8:16	14.72	562.18	--	--	--	--	--	--	--	--	NA	NA
9/12/00	9:36	14.50	562.40	--	--	--	--	--	--	--	--	NA	NA
10/5/00	11:00	14.13	562.77	--	--	--	--	--	--	--	--	NA	NA
10/13/00	11:33	14.62	562.28	--	--	--	--	--	--	--	--	NA	NA
10/31/00	12:00	15.22	561.68	--	--	--	--	--	--	--	--	NA	NA
11/27/00	11:07	13.66	563.24	--	--	--	--	--	--	--	--	NA	NA
12/1/00	12:00	12.79	564.11	--	--	--	--	--	--	--	--	NA	NA
12/8/00	11:03	13.94	562.96	--	--	--	--	--	--	--	--	NA	NA
1/3/01	10:57	11.87	565.03	--	--	--	--	--	--	--	--	NA	NA
1/30/01	12:00	12.79	564.11	--	--	--	--	--	--	--	--	NA	NA
2/1/01	16:45	11.77	565.13	--	--	--	--	--	--	--	--	NA	NA
2/13/01	11:52	11.52	565.38	--	--	--	--	--	--	--	--	NA	NA
2/28/01	12:00	11.25	565.65	--	--	--	--	--	--	--	--	NA	NA
3/20/01	9:40	8.59	568.31	--	--	--	--	--	--	--	--	NA	NA
3/20/01	15:49			7.20	569.70	7.73	569.65	--	--	--	--	20	I
3/23/01	9:05			6.29	570.61	6.79	570.59	10.93	566.45	15.65	561.73	NR	NR
4/5/01	12:37			6.86	570.04	7.39	569.99	8.45	568.93	14.38	563.00	34	I
4/6/01	11:12			6.89	570.01	7.39	569.99	8.39	568.99	14.36	563.02	34	I
4/20/01	10:30			8.28	568.62	8.78	568.60	9.82	567.56	12.06	565.32	30	I
4/26/01	12:00			--	--	9.40	567.98	9.93	567.45	11.27	566.11	NR	NR
5/1/01	9:30			9.40	567.50	9.93	567.45	--	--	11.27	566.11	30	I
5/16/01	11:56			11.04	565.86	11.56	565.82	12.18	565.20	12.48	564.90	27	I
6/4/01	10:50			9.65	567.25	10.15	567.23	10.50	566.88	10.66	566.72	27	I
6/25/01	10:44			9.64	567.26	10.15	567.23	10.38	567.00	10.47	566.91	27	I
7/10/01	10:26			11.89	565.01	12.40	564.98	12.58	564.80	12.64	564.74	26	I
7/24/01	12:30			13.50	563.40	13.99	563.39	14.12	563.26	14.26	563.12	26	I
8/13/01	10:24			13.81	563.09	14.33	563.05	14.47	562.91	14.54	562.84	25	I

Appendix 2J. Open-hole and discrete-zone manual water-level measurements in MW203R.—Continued

[Depth is depth to water, in feet. Elevation (elev.) is elevation of head, in feet above NAVD 88. "Z" means Zone in borehole. * means measurements referenced to 6-inch steel casing. All other measurements referenced to continuous multi-channel tubing. Numbers with parentheses indicates elevation of open zone in feet above North American Vertical Datum of 1988 (NAVD 88). Packer pressure in pounds force per square inch. ft, feet. NA, not applicable. NR, not recorded. I, inflated. L, low pressure; FZ, frozen. --, missing head data]

Date	Time	MW203R* Depth	MW203R* Elev.	Z1* (562.5-533.9 ft) Depth	Z1* (562.5-533.9 ft) Elev.	Z1 (562.5-533.9 ft) Depth	Z1 (562.5-533.9 ft) Elev.	Z2 (531.9-488.9 ft) Depth	Z2 (531.9-488.9 ft) Elev.	Z3 (486.9-451.9 ft) Depth	Z3 (486.9-451.9 ft) Elev.	Packer pressure	Packer condition
9/11/01	12:04	--	--	15.99	560.91	16.47	560.91	16.53	560.85	16.58	560.8	24	I
9/27/01	12:00	--	--	14.90	562.00	15.43	561.95	15.46	561.92	15.50	561.88	24	I
10/4/01	9:44	--	--	15.10	561.80	15.74	561.64	20.03	557.35	22.65	554.73	12	I
10/9/01	11:18	--	--	15.51	561.39	16.07	561.31	--	--	23.44	553.94	15	I
10/30/01	9:58	--	--	16.58	560.32	17.13	560.25	22.78	554.60	24.26	553.12	12	I
11/15/01	12:08	--	--	17.08	559.82	17.64	559.74	23.07	554.31	24.7	552.68	11	I
11/29/01	12:25	--	--	17.36	559.54	17.93	559.45	23.4	553.98	25.02	552.36	12	I
12/27/01	11:22	--	--	14.32	562.58	14.85	562.53	22.24	555.14	23.74	553.64	12	I
1/14/02	13:05	--	--	14.80	562.10	15.36	562.02	--	--	23.60	553.78	10	I
2/7/02	11:15	--	--	14.02	562.88	14.55	562.83	20.99	556.39	22.73	554.65	NR	FZ
2/28/02	13:03	--	--	13.82	563.08	14.36	563.02	17.05	560.33	21.30	556.08	5	L
3/28/02	11:04	--	--	10.85	566.05	11.35	566.03	18.92	558.46	20.37	557.01	12	I
4/16/02	12:05	--	--	11.22	565.68	11.68	565.7	17.88	559.50	19.55	557.83	11	I
5/2/02	12:14	--	--	10.57	566.33	11.07	566.31	--	--	--	--	16	I
5/7/02	12:35	--	--	10.51	566.39	11.03	566.35	17.60	559.78	19.16	558.22	16	I
6/8/02	14:35	--	--	9.88	567.02	10.42	566.96	16.40	560.98	18.53	558.85	14	I
7/10/02	13:28	--	--	12.85	564.05	13.33	564.05	20.82	556.56	20.83	556.55	13	I
8/20/02	0:00	--	--	15.43	561.47	16.02	561.36	20.90	556.48	23.99	553.39	15	I
8/20/02	19:07	--	--	15.39	561.51	15.97	561.41	--	--	22.48	--	NR	NR
9/26/02	11:30	--	--	16.26	560.64	16.83	560.55	22.79	554.59	24.97	552.41	17	I
10/28/02	12:35	--	--	14.31	562.59	14.88	562.5	22.01	555.37	23.83	553.55	15	I

Appendix 2K. Open-hole and discrete-zone manual water-level measurements in MW204R.

[Depth is depth to water, in feet. Elevation (elev.) is elevation of head, in feet above NAVD 88. "Z" means Zone in borehole. * means measurements referenced to 6-inch steel casing. All other measurements referenced to continuous multi-channel tubing. Numbers with parentheses indicates elevation of open zone in feet above North American Vertical Datum of 1988 (NAVD 88). Packer pressure in pounds force per square inch. ft, feet. NA, not applicable. NR, not recorded. I, inflated. FZ, frozen. --, missing head data]

Date	Time	MW204R*		Z1* (562.2-545.3 ft)		Z1 (562.2-545.3 ft)		Z2 (543.3-507.3 ft)		Z3 (505.3-447.6 ft)		Packer pressure	Packer condition
		Depth	Elev.	Depth	Elev.	Depth	Elev.	Depth	Elev.	Depth	Elev.		
8/9/00	9:30	6.96	568.38	--	--	--	--	--	--	--	--	NA	NA
8/14/00	10:00	7.55	567.79	--	--	--	--	--	--	--	--	NA	NA
8/17/00	12:30	7.35	567.99	--	--	--	--	--	--	--	--	NA	NA
8/21/00	9:34	8.57	566.77	--	--	--	--	--	--	--	--	NA	NA
8/22/00	10:47	7.79	567.55	--	--	--	--	--	--	--	--	NA	NA
8/30/00	9:49	8.98	566.36	--	--	--	--	--	--	--	--	NA	NA
8/30/00	12:20	8.90	566.44	--	--	--	--	--	--	--	--	NA	NA
9/11/00	8:45	9.52	565.82	--	--	--	--	--	--	--	--	NA	NA
9/12/00	11:52	9.09	566.25	--	--	--	--	--	--	--	--	NA	NA
10/3/00	10:01	8.97	566.37	--	--	--	--	--	--	--	--	NA	NA
10/10/00	17:35	9.68	565.66	--	--	--	--	--	--	--	--	NA	NA
10/31/00	12:00	10.83	564.51	--	--	--	--	--	--	--	--	NA	NA
11/27/00	11:00	7.45	567.89	--	--	--	--	--	--	--	--	NA	NA
12/1/00	12:00	7.24	568.10	--	--	--	--	--	--	--	--	NA	NA
12/8/00	10:54	8.07	567.27	--	--	--	--	--	--	--	--	NA	NA
1/3/01	10:35	5.57	569.77	--	--	--	--	--	--	--	--	NA	NA
1/30/01	12:00	6.06	569.28	--	--	--	--	--	--	--	--	NA	NA
2/13/01	11:00	5.00	570.34	--	--	--	--	--	--	--	--	NA	NA
2/28/01	12:00	4.87	570.47	--	--	--	--	--	--	--	--	NA	NA
3/23/01	10:45	2.13	573.21	--	--	--	--	--	--	--	--	NA	NA
3/23/01	13:16	--	--	1.43	573.91	1.98	573.84	1.77	574.05	1.81	574.01	18	I
4/5/01	12:00	--	--	2.78	572.56	3.29	572.53	4.89	570.93	5.39	570.43	NR	NR
4/6/01	10:52	--	--	2.84	572.50	3.38	572.44	4.80	571.02	5.25	570.57	NR	NR
4/20/01	10:12	--	--	3.84	571.50	4.32	571.50	5.67	570.15	6.09	569.73	NR	NR
5/1/01	11:30	--	--	4.55	570.79	5.07	570.75	6.26	569.56	6.40	569.42	NR	NR
5/16/01	11:29	--	--	5.80	569.54	6.33	569.49	7.28	568.54	7.31	568.51	NR	NR
6/4/01	10:28	--	--	4.36	570.98	4.49	571.33	5.40	570.42	5.44	570.38	NR	NR
6/25/01	10:24	--	--	4.75	570.59	5.26	570.56	5.61	570.21	5.62	570.20	NR	NR
7/10/01	10:02	--	--	6.72	568.62	7.25	568.57	7.55	568.27	7.57	568.25	NR	NR
7/24/01	12:05	--	--	8.65	566.69	9.17	566.65	9.44	566.38	9.47	566.35	NR	NR
8/13/01	9:45	--	--	9.01	566.33	9.54	566.28	9.87	565.95	9.88	565.94	NR	NR
9/11/01	10:55	--	--	11.90	563.44	12.38	563.44	12.62	563.2	12.64	563.18	NR	NR
9/27/01	11:35	--	--	9.61	565.73	10.16	565.66	10.47	565.35	10.49	565.33	NR	NR
10/4/01	9:10	--	--	10.66	564.68	11.21	564.61	13.29	562.53	13.3	562.52	10	I

Appendix 2K. Open-hole and discrete-zone manual water-level measurements in MW204R.—Continued

[Depth is depth to water, in feet. Elevation (elev.) is elevation of head, in feet above NAVD 88. "Z" means Zone in borehole. * means measurements referenced to 6-inch steel casing. All other measurements referenced to continuous multi-channel tubing. Numbers with parentheses indicates elevation of open zone in feet above North American Vertical Datum of 1988 (NAVD 88). Packer pressure in pounds force per square inch. ft, feet. NA, not applicable. NR, not recorded. I, inflated. FZ, frozen. --, missing head data]

Date	Time	MW204R*		Z1* (562.2-545.3 ft)		Z1 (562.2-545.3 ft)		Z2 (543.3-507.3 ft)		Z3 (505.3-447.6 ft)		Packer pressure	Packer condition
		Depth	Elev.	Depth	Elev.	Depth	Elev.	Depth	Elev.	Depth	Elev.		
10/9/01	10:52	--	--	11.41	563.93	11.99	563.83	13.66	562.16	13.7	562.12	15	I
10/30/01	9:32	--	--	13.04	562.3	13.62	562.2	14.62	561.2	14.65	561.17	11	I
11/15/01	11:43	--	--	13.64	561.7	14.21	561.61	15.02	560.8	15.05	560.77	12	I
11/29/01	11:40	--	--	13.8	561.54	14.38	561.44	15.35	560.47	15.38	560.44	9	I
12/27/01	11:34	--	--	9.79	565.55	10.35	565.47	13.11	562.71	13.13	562.69	13	I
1/14/02	12:21	--	--	9.08	566.26	9.62	566.2	12.77	563.05	12.81	563.01	13	I
1/31/02	11:56	--	--	8.98	566.36	9.50	566.32	12.85	562.97	12.89	562.93	16	I
2/28/02	11:50	--	--	8.01	567.33	8.56	567.26	11.15	564.67	11.16	564.66	13	I
3/28/02	10:16	--	--	4.31	571.03	4.80	571.02	7.53	568.29	7.57	568.25	15	I
4/18/02	10:41	--	--	4.91	570.43	5.44	570.38	8.04	567.78	8.04	567.78	12	I
5/2/02	11:50	--	--	4.06	571.28	4.58	571.24	7.30	568.52	7.35	568.47	18	I
6/18/02	13:45	--	--	4.65	570.69	5.15	570.67	7.54	568.28	7.55	568.27	16	I
7/10/02	13:18	--	--	7.26	568.08	7.74	568.08	10.52	565.3	10.5	565.32	16	I
8/21/02	10:05	--	--	11.28	564.06	11.84	563.98	13.64	562.18	13.64	562.18	NR	NR
8/21/02	10:27	--	--	11.28	564.06	11.82	564	--	--	12.61	--	NR	NR
8/21/02	14:40	--	--	11.35	563.99	11.89	563.93	--	--	13.65	562.17	NR	NR
9/26/02	11:18	--	--	12.51	562.83	13.07	562.75	14.52	561.3	14.51	561.31	18	I
10/28/02	11:58	--	--	9.2	566.14	9.74	566.08	12.67	563.15	12.66	563.16	17	I

Appendix 2L. Open-hole and discrete-zone manual water-level measurements in MW302R.

[Depth is depth to water, in feet. Elevation (elev.) is elevation of head, in feet above NAVD 88.
*, means measurements referenced to 8-inch polyvinyl chloride casing]

Date	Time	MW302R*	
		Depth	Elev.
1/3/01	11:50	20.70	554.96
1/8/01	11:10	20.83	554.83
1/30/01	12:00	20.92	554.74
2/2/01	9:14	21.01	554.65
2/13/01	12:14	20.40	555.26
2/15/01	10:50	20.33	556.83
2/26/01	9:55	20.06	555.60
2/28/01	12:00	22.18	553.48
3/15/01	12:25	19.62	557.54
8/13/01	11:07	23.80	553.36
9/6/01	11:00	24.60	552.56
9/27/01	13:19	24.68	552.48
10/9/01	12:27	24.87	552.29
10/17/01	8:44	25.02	552.14
10/18/01	9:20	25.03	552.13
10/30/01	10:55	25.37	551.79
11/15/01	10:20	25.45	551.71
11/28/01	8:16	25.48	551.68
11/29/01	10:52	25.61	551.55
12/27/01	11:13	24.31	552.85
1/15/02	13:20	23.93	553.23
2/7/02	9:45	23.13	554.03
2/21/02	13:37	22.82	554.34
3/6/02	10:16	22.28	554.88
3/28/02	12:46	21.02	556.14
4/18/02	11:17	20.62	556.54
5/7/02	14:17	20.43	556.73
5/30/02	14:19	20.25	556.91
6/18/02	11:26	20.17	556.99
7/31/02	14:00	22.75	554.41
8/22/02	13:15	24.87	552.29
9/24/02	15:53	25.11	552.05
10/28/02	14:35	23.74	553.42
11/26/02	14:45	21.38	555.78

Appendix 2M. Open-hole and discrete-zone manual water-level measurements in W202NE.

[Depth is depth to water, in feet. Elevation (elev.) is elevation of head, in feet above NAVD 88. *, means measurements referenced to 6-inch steel casing]

| Date | Time | W202NE* | |
		Depth	Elev.
2/27/01	16:07	19.77	522.38
2/28/01	15:40	19.89	522.26
4/4/01	10:30	19.01	523.14
4/10/01	10:20	19.20	522.95
4/13/01	9:50	19.14	523.01
4/13/01	12:29	19.39	522.76
4/17/01	12:00	19.51	522.64
4/19/01	11:40	19.78	522.37
4/19/01	15:11	19.60	522.55
4/26/01	9:30	19.86	522.29
4/26/01	14:24	20.07	522.08
4/26/01	18:27	21.73	520.42
5/15/01	16:23	21.10	521.05
5/17/01	17:53	18.53	523.62

Appendix 3

Manual and Continuous Water-Level Hydrographs of Boreholes in the UConn Landfill Study Area, Storrs, Connecticut

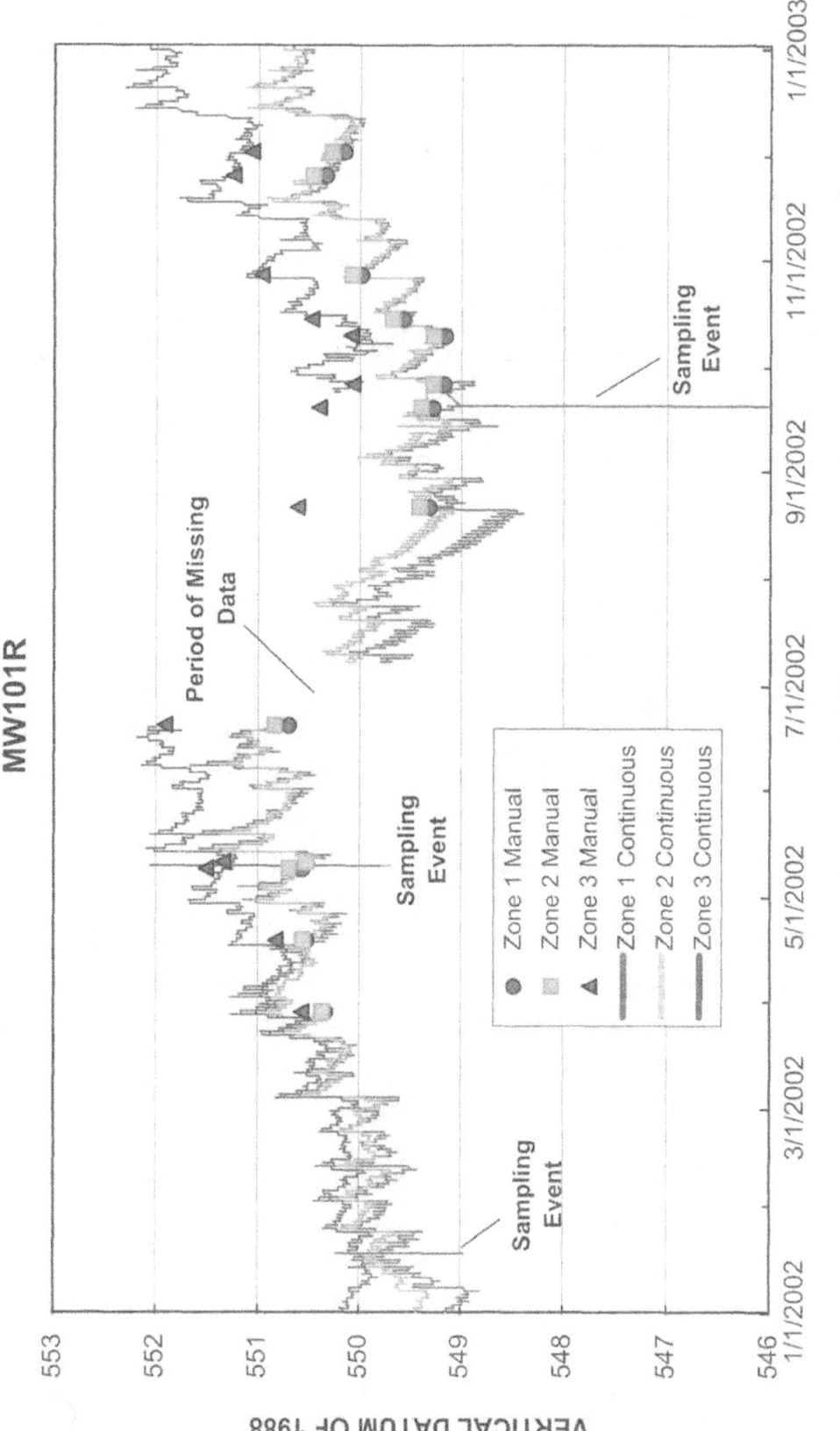

Appendix 3A. Continuous and manual water-level hydrograph of MW101R.

Appendix 3B. Continuous and manual water-level hydrograph of MW103R.

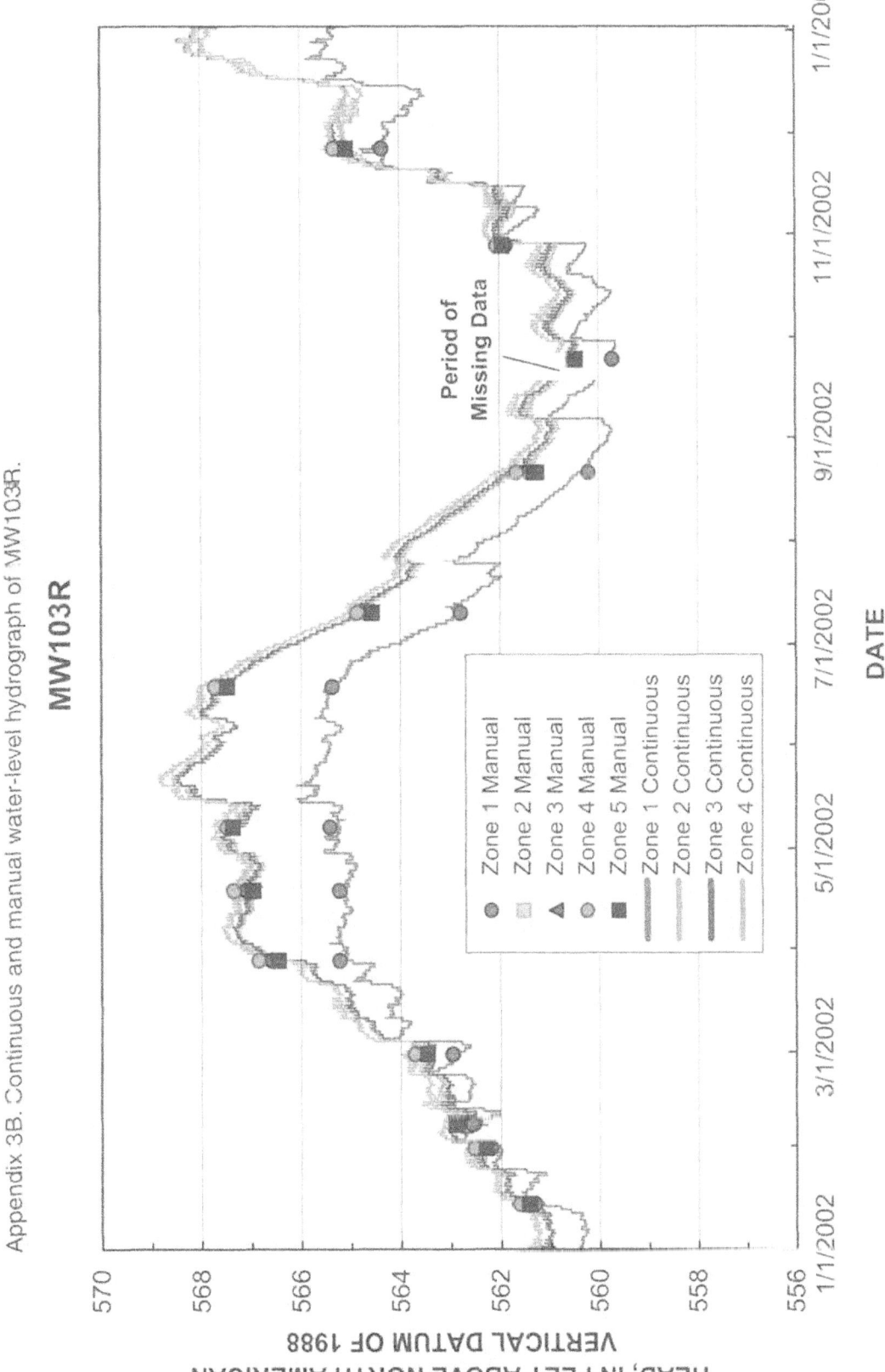

Appendix 3C. Continuous and manual water-level hydrograph of MW104R.

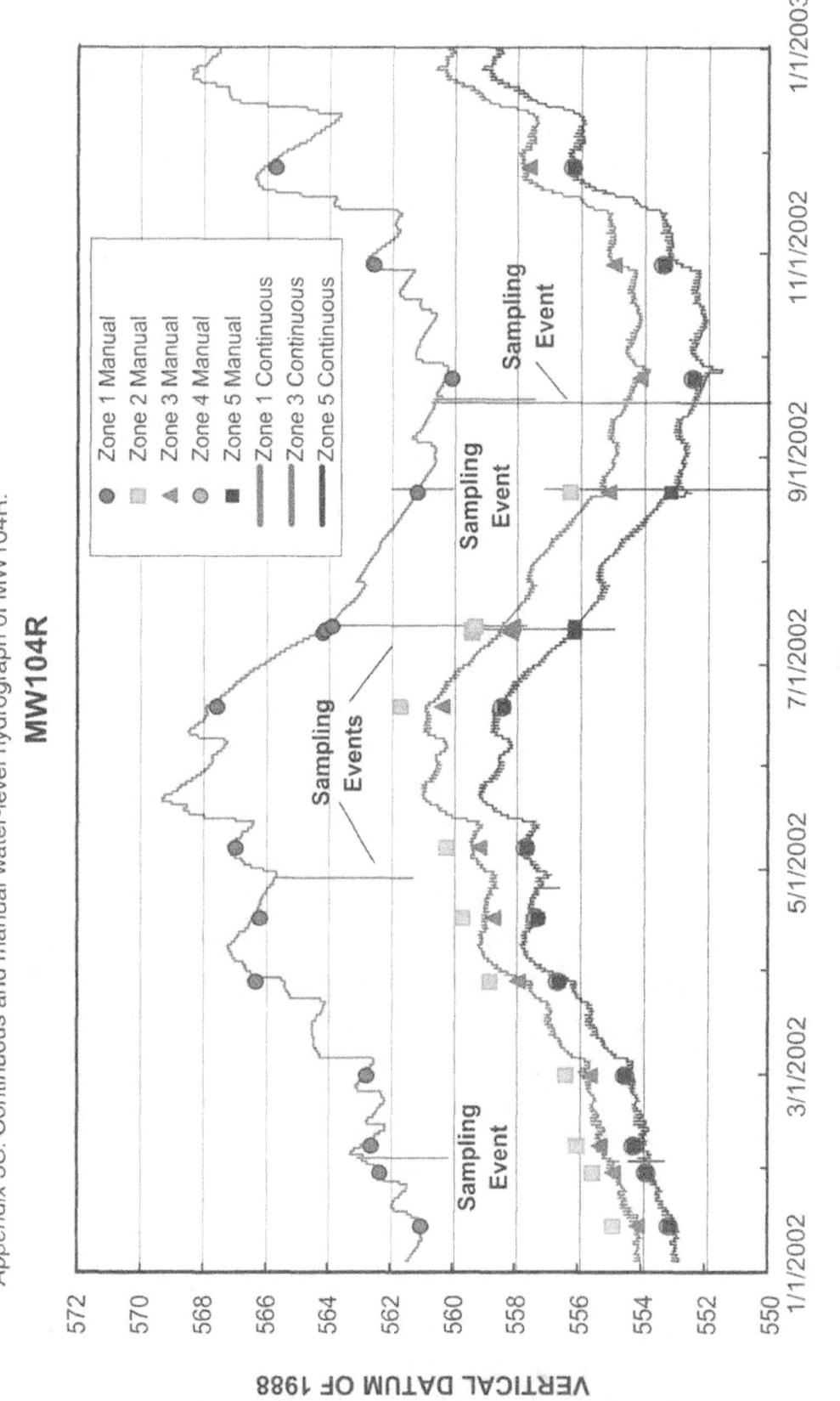

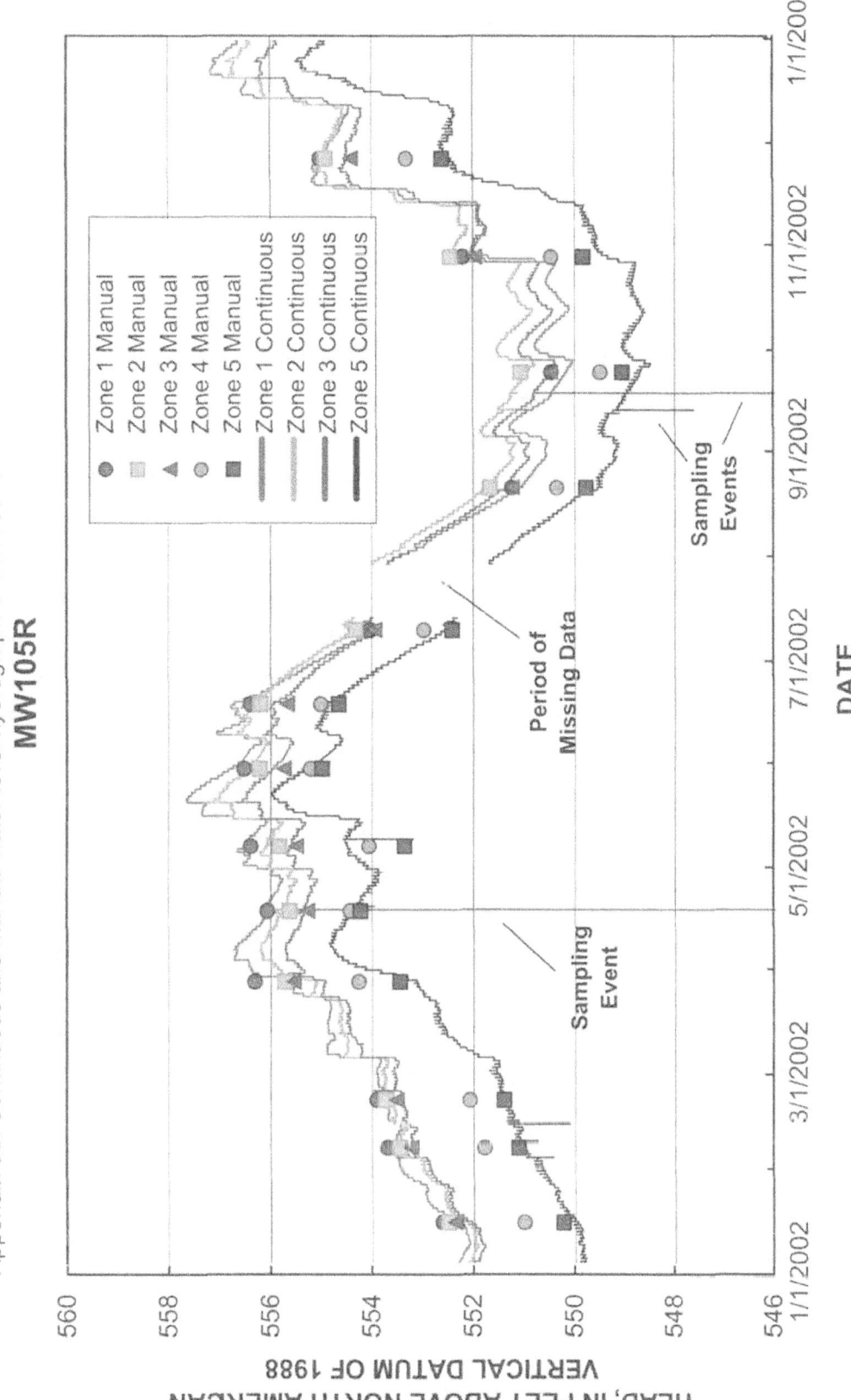

Appendix 3D. Continuous and manual water-level hydrograph of MW105R.

Appendix 3E. Continuous and manual water-level hydrograph of MW109R.

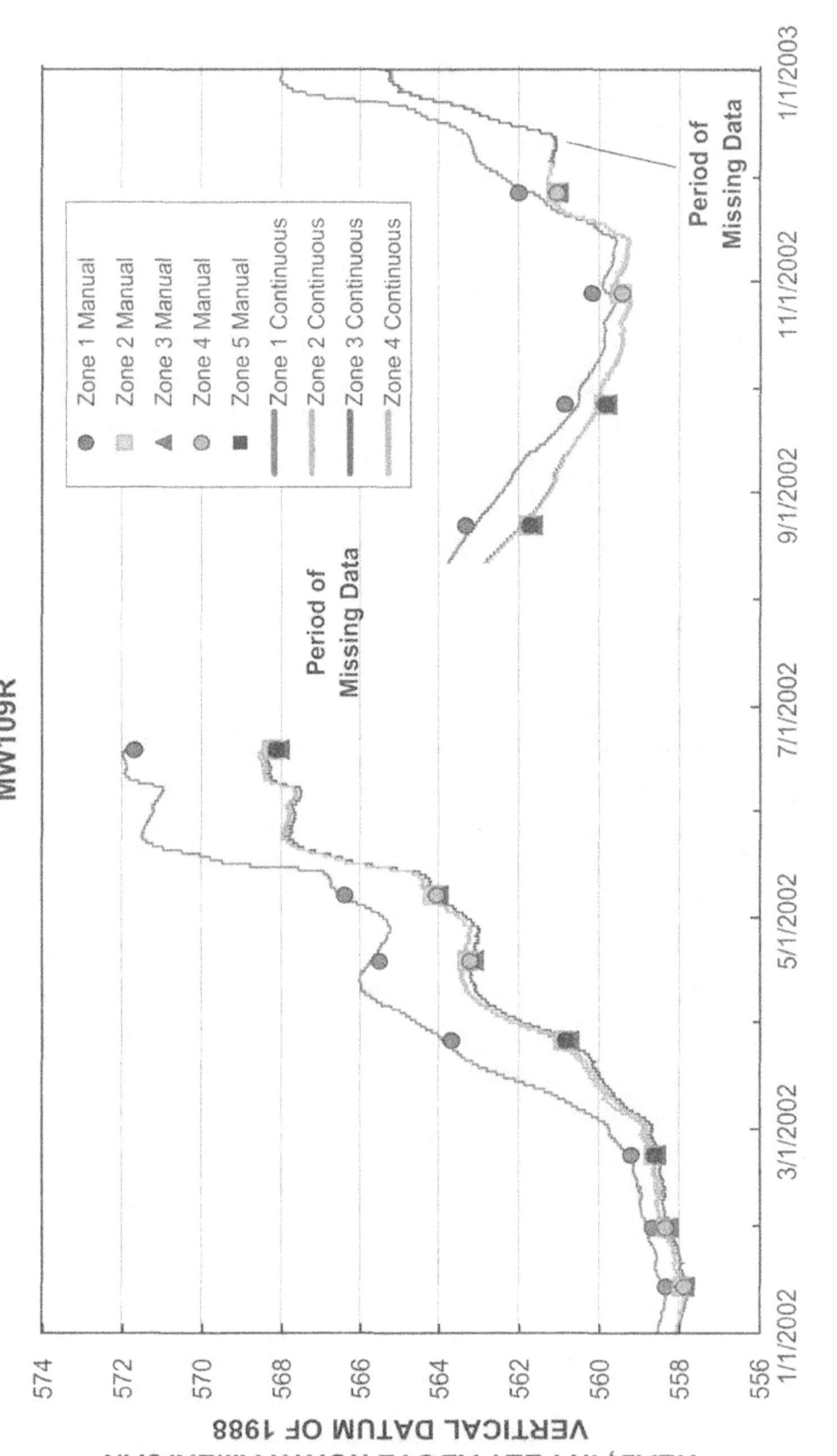

Appendix 3F. Continuous and manual water-level hydrograph of MW122R.

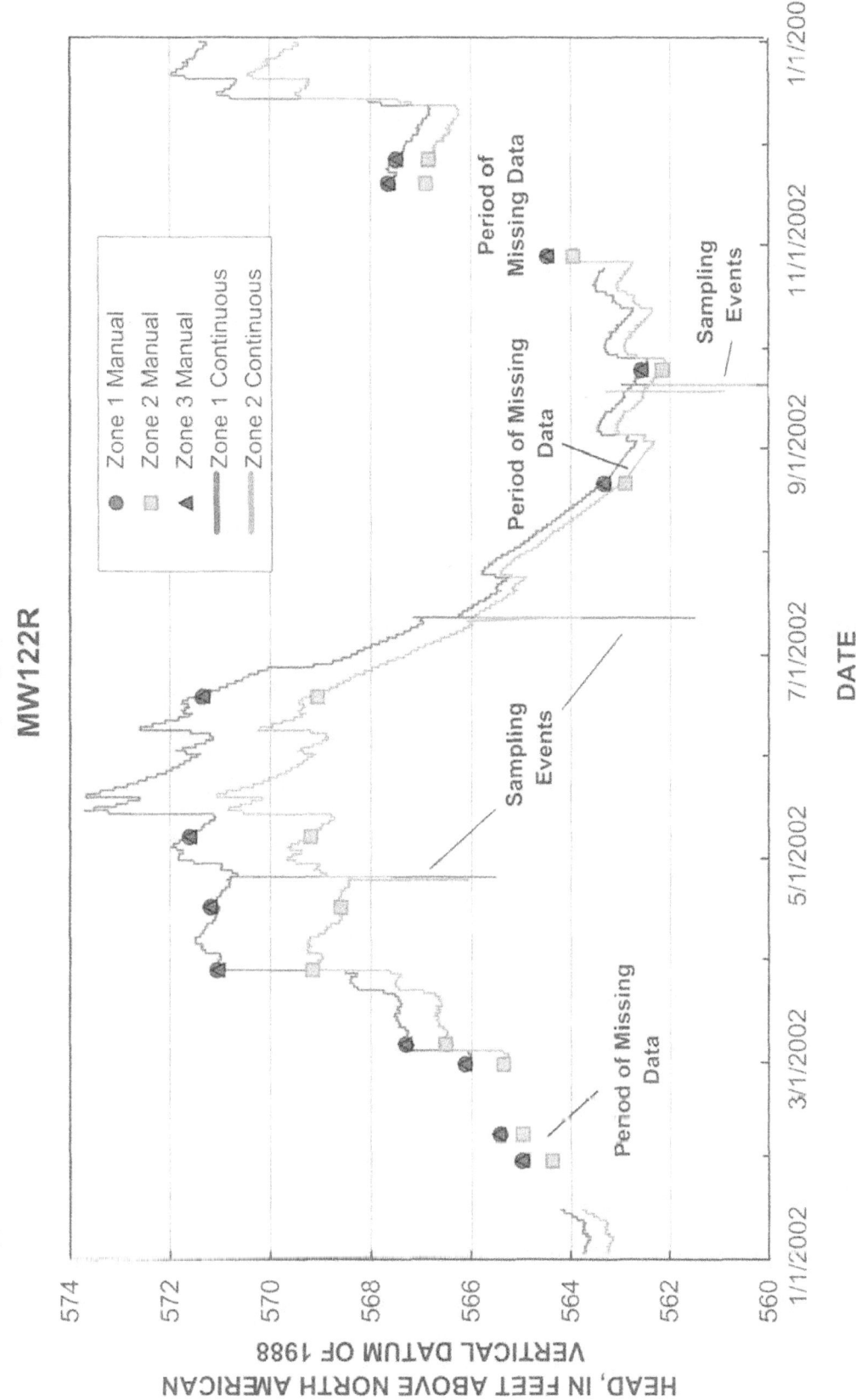

www.ingramcontent.com/pod-product-compliance
Lightning Source LLC
Chambersburg PA
CBHW081503170526
45166CB00008B/2539